AQA Science
Additional Applied

New GCSE

Gerry Blake

Jo Locke

Frances Annets

Series Editor
Lawrie Ryan

Nelson Thornes

AQA examination questions are reproduced by permission of the Assessment and Qualifications Alliance.

Published in 2011 by:
Nelson Thornes Ltd
Delta Place
27 Bath Road
CHELTENHAM
GL53 7TH
United Kingdom

11 12 13 14 15 / 10 9 8 7 6 5 4 3 2 1

A catalogue record for this book is available from the British Library

ISBN 978 1 4085 0839 8

Cover photograph: Pete Saloutos / Corbis

Illustrations include artwork drawn by Greengate Publishing and David Russell Illustration

Page make-up by Greengate Publishing

Index created by Indexing Specialists (UK) Ltd

Printed and bound in Spain by GraphyCems

Additional Applied Science

Contents

Welcome to AQA GCSE Science! 1

Working safely 2

1.1 Health and safety 2
1.2 Risk assessments 4
1.3 Fire safety 6
1.4 Following standard procedures 8
1.5 Carrying out a standard procedure 10

The use of science in maintaining health and fitness 12

Making connections 12
2.1 Your heart 14
2.2 Your lungs 16
2.3 Changes during exercise 18
2.4 Recovery after exercise 20
2.5 Controlling temperature and fluid levels 22
2.6 Physiotherapy 24
2.7 Biomechanics – the science of human movement 26
2.8 Prosthetics 28
2.9 Nutrition for exercise and fitness 30
2.10 Energy requirements 32
2.11 Sports drinks and sports diets 34
2.12 Standard procedures for maintaining health and fitness 36
Summary questions/Examination-style questions 38

The use of science to develop materials for specific purposes 42

Making connections 42
3.1 Introduction to materials science 44
3.2 Forces on materials 46
3.3 Metals and alloys 48
3.4 Composites, wood and ceramics 50
3.5 Natural or synthetic? 52
3.6 Polymers 54
3.7 Materials for sports, medicine and transport 56
3.8 Standard procedures for testing materials 58
Summary questions/Examination-style questions 60

The use of science in food production 64

Making connections 64
4.1 Introduction to food science 66
4.2 Food poisoning 68
4.3 Food hygiene 70
4.4 Standard procedures – microbiological techniques 72

4.5 Bread, beer and wine production 74
4.6 Cheese and yoghurt production 76
4.7 Growing crops 78
4.8 The use of chemicals in intensive farming 80
4.9 Rearing animals 82
4.10 The impact of intensive farming on the environment 84
4.11 Investigating plant growth 86
4.12 Rates of reaction 88
4.13 Chemical yields 90
4.14 Reversible reactions 92
4.15 Selective breeding and genetic engineering 94
4.16 Standard procedures used in food science 96
Summary questions/Examination-style questions 98

The use of science in analysis and detection 102

Making connections 102
5.1 Introduction to analytical science 104
5.2 Distinguishing different chemicals 106
5.3 Testing for ions 108
5.4 Breathalysers 110
5.5 Flame tests 112
5.6 Balanced equations 114
5.7 Titrations 116
5.8 Chromatography 118
5.9 Microscopic evidence 120
5.10 Modern analytical instruments 122
5.11 Blood and DNA 124
5.12 DNA profiling 126
5.13 Glass – the invisible evidence! 128
5.14 Standard procedures for analysis 130
Summary questions/Examination-style questions 132

How scientists used practical techniques – Assignment 1 136

6.1 Investigating the work of scientists and how they use science 136
6.2 Maximising your marks 138

How scientists used practical techniques – Assignment 2 140

7.1 How scientists use evidence to solve problems 140
7.2 Maximising your marks 142

Glossary 144

Index 148

Acknowledgements 152

Welcome to AQA GCSE Science!

This book has been written for you by the people who will be marking your exams, very experienced teachers and subject experts. It covers everything you need to know for your exams and is packed full of features to help you achieve the very best that you can.

Questions in yellow boxes check that you understand what you're learning as you go along. The answers are all within the text so if you don't know the answer, you can go back and re-read the relevant section

Figure 1 Many diagrams are as important for you to learn as the text, so make sure you revise them carefully.

Key words are highlighted in the text. You can look them up in the glossary at the back of the book if you're not sure what they mean.

 Where you see this icon, there are supporting electronic resources in our kerboodle! online service.

Learning objectives

Each topic begins with key quesitons that you should be able to answer by the end of the lesson.

 **Examiner's tip**

Hints from the examiners who will mark your exams, giving you important advice on things to remember and what to watch out for.

?? Did you know … ?

There are lots of interesting and often strange facts about science. This feature tells you about many of them.

⟳ links

Links will tell you where you can find more information about what you're learning.

Activity

Activity is linked to a main lesson and could be a discussion or task in pairs, groups or by yourself.

 Maths skills

This feature highlights the maths skills that you will need for your Science exams with short, visual explanations.

Scientists @ work

People often think of scientists wearing white coats in laboratories. However, vets, nurses, engineers, pharmacists, beauticians and gardeners are just a few careers that need scientific skills. These boxes give you an idea of how science is relevant to the workplace.

Practical

This feature helps you become familiar with key practicals. It may be a simple introduction, a reminder or the basis for a practical in the classroom.

Anything in the Higher Tier boxes must be learned by those sitting the Higher Tier exam. If you'll be taking the Foundation Tier, these boxes can be missed out.

The same is true for any other places which are marked Higher or **[H]**.

Higher

Summary questions

These questions give you the chance to test whether you have learned and understood everything in the topic. If you get any wrong, go back and have another look.

And at the end of chapters 2–5 you will find …

Summary questions

These will test you on what you have learned throughout the whole chapter, helping you to work out what you have understood and where you need to go back and revise.

AQA Examination-style questions

These questions are examples of the types of questions you will answer in your actual GCSE, so you can get lots of practice during your course.

Key points

At the end of the topic are the important points that you must remember. They can be used to help with revision and summarising your knowledge.

1.1

Health and safety

Learning objectives

- What is the Health and Safety at Work Act?

- What are the health and safety issues in science laboratories?

- Which hazard symbols and safety signs do we use in the workplace?

Figure 1 Power cables are an electrical hazard

??? Did you know ... ?

More than 1 million people have days off work each year because of problems with muscles and joints.

Half a million people are ill because of work-related stress. Although pressure keeps us motivated, excessive pressure causes illness.

Activity

The Health And Safety at Work Act

Look up the Health and Safety Executive (HSE) website for details about the Health and Safety at Work Act. Look at or print a copy of the 'Health and safety law poster pocket card'.

- What must employers do for you?

- How must you help?

Hazards and risks

Imagine that you live under power cables used by the National Grid. The high voltage could be dangerous, as there is a slight chance that a pylon could fall down in a storm and electrocute someone. In this case the risk of electrocution is small, but we have to take risks in life. There are hidden dangers in the workplace. More than 2500 people in the 16- to 24-year-old age group will be seriously injured at work this year, involving broken bones and serious burns.

> **a** List the risks you take on your way to school or college.
>
> **b** Why do you think young people face the highest risk of injury?

We cannot remove the dangers completely, but we can reduce the risks of those dangers hurting us.

Think about the main reasons why people get injured:

- Slips and trips – 33% of all major injuries. The cost to employers is £512 million and the cost to the heath service is £133 million, per year. The human cost cannot be calculated.

- Unsafe lifting and carrying, causing back pain.

- Falling from a height – causing 70 deaths and 4000 serious injuries each year.

- Being struck by a moving object. Each year, 3500 people are killed on our roads and 40 000 are seriously injured.

The **Health and Safety at Work Act** of 1974 was put in place partly so employers and workers made provision for securing the health, safety and welfare of people at work. It also helps to protect the public against risks to health or safety in connection with the activities of people at work. The **Health and Safety Executive** (HSE) is responsible for the regulation of risks to health and safety in the workplace.

In scientific workplaces, there are hazardous substances and dangerous equipment to be dealt with. There are also electrical risks and noise and manual handling problems. Scientific work can be dangerous.

The job of the Health and Safety Executive (HSE) is to try to control the risks to people's health and safety at work. But working in a laboratory is like driving a car – you are responsible for your own safety. You cannot rely on the efforts of car manufacturers, driving instructors or other road users to protect you. Similarly you must not expect that other people will always keep you safe in a scientific workplace.

> **c** Why do you think accidents among scientists are rare?

Hazard symbols and safety signs

- Yellow and black triangles *warn of danger*, e.g. radioactivity.

- A red crossbar tells you what you *must not do*, e.g. no smoking.

- A blue circle tells you something you *must do*, e.g. wear eye protection.

- A green background gives *safety information*, e.g. first aid.

- *Hazard symbols* are orange and black, e.g. highly **flammable**.

Figure 2 Some common safety signs

The symbols in Figure 2 are used in the workplace. You will see the hazard symbols in Figure 3 on chemicals in the science lab.

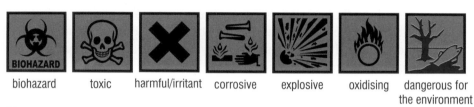

Figure 3 Hazard symbols

You already know some risks of using poor experimental technique. Health and safety regulations, if followed, protect you from danger.

The school laboratory technicians have to take care when they prepare solutions. For example, solid sodium hydroxide is corrosive, causing severe burns. It is particularly dangerous to eyes. It gets hot when added to water.

Sodium hydroxide solution is an alkali; sometimes you will use it in science experiments. Your school technician prepares your solution by dissolving solid sodium hydroxide in water.

d When preparing solutions of sodium hydroxide, what safety precautions should your school technician take?

e Think about the safety rules in your school or college laboratory. What five rules do you think are the most important and why?

Summary questions

1 a Which of the following mandatory and safe condition signs corresponds to:
  i wear hand protection
  ii wear breathing mask
  iii wear ear protection
  iv eye wash
  v emergency shower
  vi fire alarm point?

b Which colour is associated with:
  i a fire sign
  ii a mandatory (must do) sign
  iii a safe condition sign?

Activity

Lab safety

1 Find out about safety precautions taken in your laboratories to keep you protected. Consider mains electricity, gas, toxic fumes, fire and explosions.

2 Using secondary sources (books or the internet) find out what these words mean:
  a highly flammable
  b biohazard
  c toxic
  d harmful
  e irritant
  f explosive
  g oxidising.

Key points

- The Health and Safety at Work Act puts the responsibility for safety on both employers and employees.

- Biological, chemical and physical hazards exist in science labs.

- Hazard symbols and safety signs are easy to recognise once you know what they look like.

1.2 Risk assessments

Taking risks

Learning objectives

- What is the difference between hazards and risks?
- Why do we need to make risk assessments?

Some people like taking risks, e.g. whilst skiing or bungee jumping.

Sometimes the hazards come with the job, like being a policeman or a fireman.

Sometimes we are not aware there is a risk.

A **hazard** is anything that can cause harm. These include chemicals, electricity and gas, excessive noise, careless behaviour, microorganisms, etc. The **risk** indicates the chance that someone will be harmed by the hazard and the seriousness of the consequences.

   a What is the difference between a hazard and a risk?

Each year 250 people lose their lives at work in Britain. More than 150 000 are injured or badly hurt.

Did you know ... ?

There are 1.6 million injuries at work each year.

70% of these could be prevented if employers put control measures in place.

The injury rate in firms with fewer than 50 employees is over twice the rate of that in firms employing more than 1000 people.

Figure 1 Safety slogans

In your school or college laboratory, the hazard caused by touching hot tripods and test tubes is fairly common. We could assess that the risk of a burn occurring would be high, but the consequences of the hazard are usually not very serious. Control measures (or safety precautions) can be used to reduce the risks of a burn.

   b What control measures would you use to reduce the number of minor burns in schools and colleges?

   c What is the emergency action (first aid) for a minor burn accident?

An electric shock from a faulty appliance could result in death. This is a major hazard. However, with properly designed and maintained equipment, the risk is insignificant.

Concentrated sulfuric acid is corrosive. It causes severe burns and can permanently damage eyes. It is a major hazard. If no control measures are put in place and concentrated sulfuric acid is used in an experiment in the laboratory, the risk of permanently damaging eyes or the acid causing severe burns is significant.

> **d** What control measures can we use to reduce the risk when using sulfuric acid?

Writing risk assessments

We make **risk assessments** to ensure that we control hazards and limit risk.

> **e** Explain why the probability of being harmed by animals at a zoo is relatively small, but the consequences of injury could be very high.

In a risk assessment we:

1 Identify **hazards**, i.e. things that could cause harm – materials, unsafe procedures and equipment.
2 Work out the **risk** – how likely is it that harm could occur and how serious could the consequences be?
3 Put **control measures** (safety precautions) in place to avoid or reduce the risk as far as possible.
4 Decide on the **emergency action** (first aid) to take if the controls fail and there is an accident.

The table below shows how to set out a risk assessment:

Risk Assessment Form

Name of student: _____ Date: _____

Task: _____

Hazards	Risks	Control measures	Emergency action
Think about the material, procedure or equipment used. What makes it dangerous and what could go wrong?	Probability of harm: likely/ possible/unlikely Seriousness of consequence: extremely harmful/harmful/ slighty harmful	Safety precautions	First aid

Table 1

Health and safety checking

Always examine the area you are working in and reduce the risk to yourself and others by:

● Knowing the person to contact concerning any potential hazards – your safety representative.
● Checking the layout of your work space, safety information and working policies. Identify the potential hazards and reduce the risks.
● Following risk assessments, carrying out tasks safely, staying alert to problems and reporting any incidents that arise.

Figure 2 An unexpected hazard

⧉ links

Please refer to Chapter 7 when completing your Assignment 2 investigation and refer to Table 1 when writing your risk assessment.

Summary questions

1 Explain why each part of a risk assessment form matters as regards:
   a hazard
   b risk
   c control measures
   d emergency action.
2 Think about a workplace you have been to recently. Suggest how the phrase 'Think risk, assess, reduce, avoid' applies to someone working there.

Key points

● A hazard is a danger; a risk is the chance of being harmed.

● A risk assessment helps you to control hazards and limit risks.

1.3 Fire safety

Preventing fires

You are twice as likely to die in a fire in a house that has no smoke alarm than one that has. Smoke is the main cause of death and damage to property in fires. Each year about 700 people die from fires in their own homes and 14 000 are injured. 60% of home fires result from cooking. Only 6% of deaths and 10% of injuries happen in workplace fires.

a Why is it important to check your smoke alarm regularly by pressing the test button?

For a fire to burn, it needs three things to be present:

OXYGEN + HEAT + FUEL = FIRE

A fire is prevented or extinguished by removing any one of them.

b Why should you 'stop, drop and roll' even if there is a fire blanket?

Figure 1 The fire triangle

Figure 3 What should you do if your clothes catch fire?

c Explain how a heavy metal fire door removes oxygen, heat and fuel.

Fire extinguishers

Water

Water fire extinguishers work by cooling material to below its ignition point. Water can be used for burning solids such as paper, wood, plastic, etc. A water extinguisher is not suitable for liquids that float on water such as paraffin, petrol, oil, or for fires involving electrical apparatus.

Dry powder

Dry powder extinguishers work by smothering the fire (by excluding oxygen) and knocking down flames. They can be used for burning solids, liquids or for electrical fires. They are best for fires involving flammable liquids. It can be dangerous to extinguish a gas fire without first turning off the gas supply.

Figure 2 Smoke reaching a smoke alarm on the ceiling

Figure 4 Fire doors form a barrier to stop fire spreading. They must be kept shut but not locked!

Water	Powder	Foam	Carbon dioxide (CO_2)
For wood, paper, textiles and solid material fires	For liquid and electrical fires	For use on liquid fires	For liquid and electrical fires
DO NOT USE on liquid, electrical or metal fires	**DO NOT USE** on metal fires	**DO NOT USE** on electrical or metal fires	**DO NOT USE** on metal fires

Figure 5 Types of fire extinguisher

Foam

Foam extinguishers work by forming a foam blanket to smother the fire and stop combustion. They work well for burning solids and liquids. Foam extinguishers are not recommended for home use or electrical fires.

Carbon dioxide (CO_2)

Carbon dioxide fire extinguishers smother the fire by displacing oxygen. They are ideal for fires involving electrical equipment, and will extinguish flammable liquid fires. Carbon dioxide can be dangerous in confined areas as it prevents you breathing.

d   In electrical fires, why could water be dangerous?

e   Which extinguisher is best for electrical fires?

f   Which is the only type of extinguisher that removes heat from a fire?

AQA Examiner's tip

In your coursework report include sections about fire safety in your:
- school or college
- chosen specific workplace.

Summary questions

1   Copy and complete the sentences using the words below:

electricity   exit   alarm   windows   gas   doors

When you find a fire, sound the .... Shut .... and .... If you can, switch off the .... and .... Then make your way to the nearest ....

2   a   Why is water dangerous on oil fires?
    b   Why is using a fire blanket better than using an extinguisher when a chip pan catches fire?
    c   Powder on upholstery can put out a fire, only for it to relight. Why?
    d   Suggest why foam extinguishers are not suitable for use in homes.
    e   Why is carbon dioxide (CO_2) dangerous in confined areas?

Key points

- Fire safety prevents severe burns, loss of life and damage to property.
- You need to know what types of fire should be extinguished with water, powder, foam and carbon dioxide extinguishers.

1.4

Following standard procedures

Learning objectives

- What is a standard procedure?
- Why do we need standard procedures?
- What is the scientific method?

⊂∪⊃ **links**

For more information on standard procedures in healthcare see 2.12 Standard procedures for maintaining health and fitness; on materials see 3.8 Standard procedures for testing materials; on food see 4.16 Standard procedures used in food science; and for analytical sciences see 5.14 Standard procedures for analysis.

Figure 1 Nurse cleaning an incubator

Recipes

When baking a cake, it's no good just putting flour, eggs and a few other bits and pieces into a pan and sticking it in the oven. You need certain amounts of each ingredient that you mix and cook in a certain way. It's better to follow a step-by-step recipe.

Scientists have to follow recipes, too. They're called 'standard procedures', and you'll find lots of them in the workplace. These precise instructions tell you how to carry out tasks, how to make things and how to test things. A standard procedure is an agreed way of doing something.

a How is a standard procedure like a recipe?

Scientists follow standard procedures to make sure the materials and products they investigate, or make, are safe and reliable in use. For example, public health laboratories test food and drink samples in specific ways. There are British Standards for analysing ice cream, testing water quality, and determining the bacterial content of milk. Instructions in standard procedures are very precise.

b Why do we need standard procedures? (Think about a nurse cleaning an incubator.)

If you're making or testing something using a British Standard, you do it in exactly the same way as someone else making or testing the same product. If someone did something differently, the product or test results would be different.

c Why do standard procedures need to have so much detailed information?

The scientific method

Scientists the world over take a similar approach to asking and answering real-life questions. Scientists call this way of researching the **scientific method**. Think of it as a series of steps:

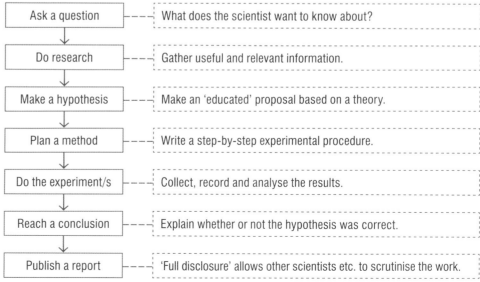

Figure 2 The scientific method

When carrying out experimental work, scientists use standard procedures. This helps to cut out mistakes and allows other scientists to perform the tests again. Indeed, anyone seeing a scientific research report should be able to **repeat** the steps taken for themselves and **reproduce** the results obtained.

Close examination or scrutiny of a scientific report may show up possible sources of **error**. Then, as a result of improving techniques, different conclusions may be reached. In this way science is an ongoing process.

d What is a hypothesis?

e Explain the sentence, 'Full disclosure allows others to scrutinise the work'.

Activity

An analytical scientist in the making!

Dave watches his mum bake bread. He asks her what makes the bread rise. She explains, 'Yeast releases a gas as it feeds on sugar'. Dave wonders if the amount of sugar affects the size of the bread loaf. He researches into baking and fermentation and comes up with a safe way to test his hypothesis that 'If more sugar is added, then the bread will rise higher because the yeast will have more food so will produce more carbon dioxide'. Dave writes out the procedure for his experiment, based on his mum's recipe. After carrying out his experiment and taking results, he examines his data. Dave rejects his original hypothesis, concluding instead that a certain volume of sugar (not necessarily the largest volume) produces the largest loaf. He tells his mum about his findings and prepares a poster of his project.

● Explain how Dave is using both a standard procedure and the scientific method in his investigation.

Summary questions

1 How is the flow chart in Figure 3 like/unlike a standard procedure?

Figure 3 Getting out of bed in the morning

2 How does a standard procedure help you get reproducible results?

Key points

● A standard procedure is an agreed way of doing something.

● Scientists follow standard procedures to obtain consistent and reproducible results.

● The scientific method is a set way of carrying out scientific research.

1.5

Carrying out a standard procedure

Learning objectives

● How do scientists follow a standard procedure?

Figure 1 Lab safety – the only way is the safe way

∞ links

For information on the importance of standard procedures look back to 1.4 Following standard procedures.
For more information on your Assignment 1 investigation see Chapter 6.

Introduction

When monitoring and controlling processes, and when making and analysing substances, scientists need results they can depend on. Standard procedures ensure everyone carries out a particular task in exactly the same way. This helps to make their results repeatable and reproducible.

● Before you start an experiment always read the procedure carefully, checking to see if there is anything you do not understand.

● Set out your work area safely and tidily with the equipment and materials you need.

● During the experiment, carry out the instructions safely and carefully, one step at a time.

● Be alert to possible sources of error, and repeat observations, which should then be reproducible by others who follow the same procedure.

a Suggest a slogan to encourage laboratory safety.

Activity

Testing materials

You are a materials scientist working for a sportswear manufacturer. Your employer wants you to assess the qualities of two plastics being considered for the hard plastic outer shell of a bicycle helmet. You must compare their hardness by following the British Standard test for impact resistance (BS EN ISO 6603-1:1996) described on the next page. Your report must conclude with a recommendation about which plastic to use, based on the relative merits of the two materials.

Figure 2 Manufacturer's bicycle helmet

Practical

Standard procedure – impact resistance

Comparing the impact resistance of materials

1 **Scope:** This standard procedure is adapted from the 'Falling dart method' to compare the impact resistance of sheets of various materials (such as plastics, wood, cardboard, ceramic tiles and composites).

2 **Definitions:** *Impact resistance* – the force needed to break, crack or dent the specimen.

Impact-failure energy = weight dropped × height dropped.

$$= \text{mass (kg)} \times 10\,\text{N/kg} \times 0.5\,\text{m}$$
$$= (\text{mass (g)} \div 1000) \times 10 \times 0.5$$
$$= \text{mass (g)} \times 5 \div 1000$$
$$= \text{mass (g)} \div 200$$

3 **Principle:** Apply an impact force by dropping a weight onto a ball-bearing resting on the material. Increase the impact energy by increasing the weight dropped, until the specimen breaks.

4 **Apparatus:** Test materials, 100 g slotted masses set on a hanger, 0.5 m long guide tube, clamp stand, petroleum jelly-coated ball-bearing.

5 **Test specimens:** This is only a fair test if the test specimens have the same thickness. At least three specimens of each material should be tested.

6 **Procedure:**

Set up a testing rig as shown in Figure 3.

Drop the 100 g mass hanger from a height of 0.5 m onto the ball-bearing.

Remove the ball-bearing and record changes, if any, to the test specimen (e.g. size of cracks and dents).

Put the ball-bearing back into position.

Add a 100 g mass to the hanger and drop it onto the ball-bearing again.

Record any changes.

Continue repeating this procedure, adding 100 g masses (up to 1 kg). Stop if the specimen breaks.

7 **Safety:** Wear eye protection. Protect bench top. Avoid falling masses and bits from fractured materials.

8 **Expression of results:** Calculate the impact-failure energy (in joules) using the equation found in point **2** above:

$$E \text{ (in J)} = \text{mass (in g)} \div 200$$

9 **Test report:** Write a report on your findings, this should be well presented so others can easily see your evidence and conclusion. Your test report should include:

a reference to this standard procedure

b the identity (and thickness) of each test specimen

c the nature of cracks and dents produced in each specimen

d the average impact-failure energy of each test specimen

e a recommendation (your conclusion) about the suitability of each of the two materials for the plastic shell of a bicycle helmet.

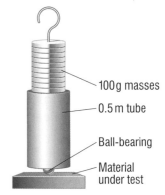

Figure 3 Impact resistance

- 100 g masses
- 0.5 m tube
- Ball-bearing
- Material under test

Summary questions

1 When comparing two materials experimentally, why must they both be the same size as each other?

2 Write a risk assessment form for the standard procedure to compare the impact resistance of materials.

Key points

When following a standard procedure, scientists:

- First check they understand it.
- Set out their work area safely and tidily.
- Follow the instructions step-by-step.
- Keep alert to sources of error and repeat observations.

Making connections

In this chapter you will learn about some of the science and techniques used by healthcare scientists. You will look at the different occupations of healthcare scientists, the effects of exercise on the human body, health and nutrition, and how health scientists deal with injuries to muscles and the skeleton.

Figure 2 Physiology: Sports physiologists are employed to maximise the performance of an athlete's body.

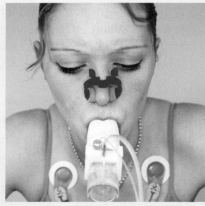

Figure 3 Energy for exercise comes from aerobic respiration (or anaerobic respiration if there is a lack of oxygen).

Figure 7 Doctors: We rely on the expertise of our health service 'from the cradle to the grave'.

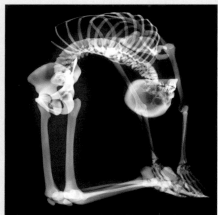

Figure 1 Healthcare scientists not only play a huge role in society, but also for athletes.

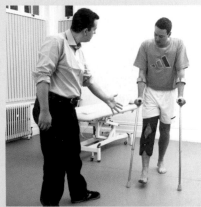

Figure 6 Physiotherapy: Physiotherapists use exercise to help their patients regain mobility and strength after an injury or surgery.

Figure 4 Fitness training increases the response of the heart, lungs and muscles. Improvements can be assessed by taking baseline measurements.

Figure 5 Appropriate diets take into account the nature of the sport and the personal daily energy requirement of the athlete who is competing.

Healthcare professionals

Healthcare science is a partnership between those who work in health and fitness.

Doctors, nurses and pharmacists help prevent, diagnose and treat a huge number of medical conditions.

Nutritionists and dieticians encourage us to stay fit and healthy. Athletes rely on their expertise. They can improve performances by following professional advice about controlling energy and nutrient intake.

Health and fitness are important to physiologists. They need to deal with those parts of the body involved in sport and exercise activities. Sports scientists and training coaches often study the physiological changes of an athlete during intense training. They use the results to advise the athlete about maintaining personal fitness.

Physiotherapists specialise in treating physical problems caused by accidents, illness and ageing, particularly those affecting muscles.

a   State four occupations in healthcare science.

b   Describe the role of two of these healthcare scientists.

c   Describe the role of any fitness practitioner.

d   Imagine you are skilled in a particular sport, yet today you are not successful. Explain why your body may be under-performing.

Activity

Healthcare scientists

Produce a slide for a PowerPoint presentation on the work of a particular healthcare scientist. Focus on **one** of the following:

● their specific role
● how their work affects health and fitness
● how their work is applied to sports men and women.

Combine your slides into one class slideshow.

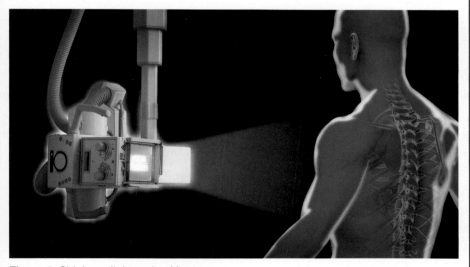

Figure 8 Shining a light on healthcare

Scientists @ work

Search the internet for careers in healthcare, or go to:
www.connexions-direct.com
www.health-care-careers.org/
Try the career quiz on:
www.teentoday.co.uk/quizzes/careerquiz

Figure 9 Questioning your future

2.1

Your heart ⓚ

Learning objectives

- What is the structure of the heart and its blood vessels (cardiovascular system)?

- How do you measure your heart rate (pulse)?

Feel your ribs. They protect your heart and lungs (see Figure 1).

Your heart is about the same size as your clenched fist. It is a **double pump** made of cardiac muscle that pumps blood round your body.

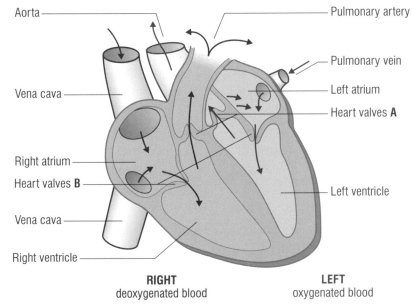

Figure 2 The heart has four chambers

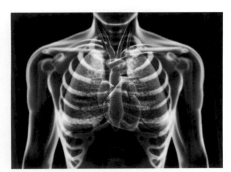

Figure 1 The human thorax or chest

- The **right atrium** receives deoxygenated (oxygen-poor, carbon dioxide-rich) blood from the organs of the body.
- The **right ventricle** pumps this deoxygenated blood to the lungs.
- The **left atrium** receives oxygenated (oxygen-rich, carbon dioxide-poor) blood from the lungs.
- The **left ventricle** pumps this oxygenated blood out to all the organs (except the lungs).

  a Do all veins carry deoxygenated blood?

  b What are the lower chambers of the heart called?

When the heart beats:
- The atria gently push the blood into the ventricles. Valves B open and valves A close.
- Then the ventricles contract, forcing blood down the arteries. Valves A open and valves B close. The valves only allow the blood to pass through the heart in one direction.

During exercise your physiology changes, i.e. your body operates differently. Your heart rate increases and more blood is pumped round your body with each beat.

??? Did you know ...?

In women, the heart volume is usually about 0.5–0.6 litres. Some female swimmers have volumes of 0.9 litres.

An elite male cyclist's heart volume could be near 1.7 litres.

AQA Examiner's tip

Remember:
- Blood leaves your heart through your **arteries**.
- Blood returns to your heart through your **veins**.
- Your pulmonary vein carries oxygenated blood *from* your lungs *to* your heart.

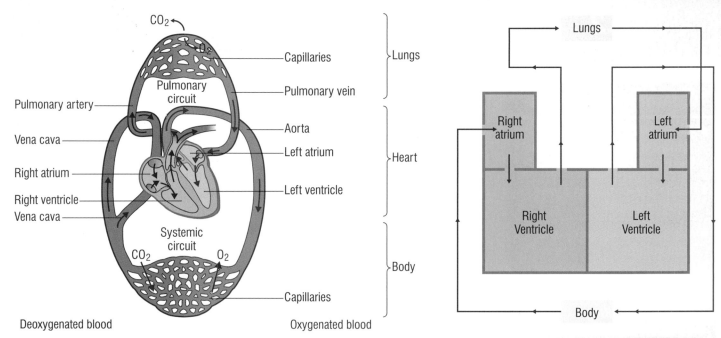

Figure 3 Blood circulation

Activity

Heart rate (pulse)

Healthcare scientists advise on personal fitness once they have taken **baseline measurements** before, during and after exercise.

1 After sitting quietly for five minutes, measure your heart rate. This is your **resting heart rate**. Either use a pulse meter or find your pulse by pressing an artery against a bone (on your neck or at your wrist). Count the pressure pulses for 30 seconds. The average resting heart rate is between 60–80 beats per minute.

2 Now do about one minute of step-ups or skipping and measure your pulse just before you do one of these exercises.

3 Immediately after you finish the exercise, record your heart rate again. Copy and fill in a table of results:

Resting heart rate (beats per minute)	Pre-test heart rate (beats per minute)	Post-test heart rate (beats per minute)

Safety: Only do this physical activity if you are fit and healthy.

● What may cause your pre-test heart rate to change?

● Why is your post-test heart rate greater than your resting heart rate?

● Suggest some everyday factors that may cause your normal heart rate to vary.

Summary questions

1 How many chambers does the human heart have?

2 Why is the heart called a 'double pump'?

3 **a** The left ventricle has thick walls and is stronger than the right ventricle. Why do you think this is?

   b Why are ventricles thicker-walled than atria (the plural of atrium)? (**Hint:** Think where the blood goes.)

4 Some people have larger hearts than others. How does this help in sports?

Key points

● Your cardiovascular system is your heart and its blood vessels.

● Blood leaves your heart's ventricles through arteries. Blood returns to your heart's atria through veins.

● Your right ventricle pumps deoxygenated blood to your lungs. Your left ventricle pumps oxygenated blood to the rest of your body.

2.2 Your lungs k

Learning objectives

- How does the structure of the thorax enable ventilation of the lungs?
- How do you measure your breathing rate?
- What is vital capacity and tidal volume?

?? Did you know ...?

You have about 300 million alveoli, with a **surface area** equivalent to half a tennis court!

AQA Examiner's tip

Remember when you breathe in your lungs increase in size because your:
- diaphragm contracts and moves down
- intercostal or rib muscles contract and lift your ribs up and out.

Your lungs are like big sponges that exchange the gases oxygen and carbon dioxide. At the ends of the bronchioles are air sacs, called **alveoli** (see Figure 1). The alveoli have thin walls and are surrounded by blood capillaries. This allows for efficient gas exchange in and out of the blood capillaries. Oxygen from the air passes into the red blood cells, at the same time as carbon dioxide leaves. The **deoxygenated blood** that arrived at the lungs in the pulmonary arteries (blue in Figure 3 on the previous page), now returns to the heart as **oxygenated blood** through the pulmonary vein (red in Figure 3 on the previous page).

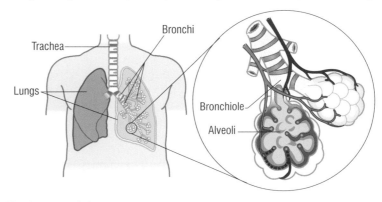

Figure 1 The lungs and air sacs

Breathe in and out. Feel your sternum (breastbone) and your diaphragm (see Figure 3).

As the **muscles** in your **thorax** relax, your lungs decrease in size, and you breathe out.

When breathing in, the volume of your lungs increases because:
- the **diaphragm** contracts and moves down,
- the **intercostal muscles**, between your ribs, contract and pull your ribcage and sternum up and out.
- This lowers the pressure in your lungs so air rushes in from higher pressure air around you.

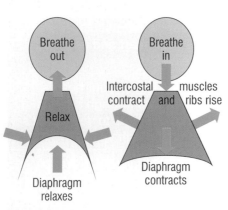

Figure 2 Breathing in increases the size of your thorax whereas breathing out decreases its size

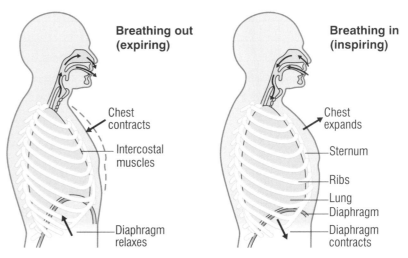

Figure 3 Breathing out and breathing in

a Where in the lungs does gas exchange take place?

b What does your diaphragm do to inflate your lungs?

c Where are the intercostal muscles?

d How do your intercostal muscles help to ventilate (force air into) your lungs?

e Breathing **ventilates** your lungs with fresh air containing 21% oxygen. You exhale air containing only 16% oxygen. How much extra carbon dioxide is breathed out?

Practical

Measuring lung volumes with a digital spirometer

Safety: Take appropriate precautions and wear suitable clothing.

Take the following measurements before and after exercise, such as a 400 m race:

1 **Breathing rate** in breaths per minute (counting for one minute). The average resting breathing rate is 12 breaths per minute.

2 **Tidal volume (TV)**. TV is the volume of air you breathe in and out normally. Measure this volume in cm^3 using a spirometer.

3 **Vital capacity (VC)**. VC is the maximum volume of air you can force out after breathing in as hard as you can. (It is the volume you expire after maximum inspiration.)

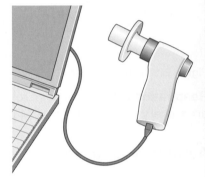

Figure 4 USB spirometer

Figure 5 Tae Kwon Do athletes breathe into masks to measure their oxygen consumption as they exercise

During exercise, your breathing rate increases. There is not much difference between the lung capacities of athletes and non-athletes. However, the amount of air that an athlete can breathe in and out in one breath (the tidal volume) does increase slightly with regular exercise. This is due to the intercostal muscles becoming stronger.

f How do your lungs put more oxygen into your blood during exercise?

Summary questions

1 What is the difference between inspire and expire?

2 What two muscles help you to breathe in?

3 How does the large surface area of your lungs help when breathing?

4 How do your lungs and heart work together to get oxygen to your muscles?

5 How does having a large tidal volume help an athlete?

6 Why is more carbon dioxide breathed out than breathed in?

Key points

● Your diaphragm and the intercostal muscles between your ribs change the air pressure in your lungs to make you breathe.

● If these muscles in your thorax relax, your lungs decrease in size and you breathe out.

● When you breathe in your lungs increase in size because:

– your diaphragm contracts and moves down

– your intercostals contract and lift your ribs up and out.

2.3 Changes during exercise

Learning objectives

- How do the heart and lungs help to provide glucose and oxygen to the muscles?

- How do aerobic and anaerobic respiration differ?

- What changes occur to breathing and heart rate during exercise?

AQA *Examiner's tip*

Respiration is *not* breathing in and out. Respiration is the *process* of releasing energy from glucose. You only need the symbol equation for aerobic respiration in the Higher Tier exam. In your blood, plasma carries glucose and red blood cells carry oxygen.

Respiration

We all need a diet that provides us with enough energy to live. **Carbohydrates** give us energy. We get carbohydrates from **sugars** and **starch**. Bread, pasta, rice and potatoes contain starch.

As you digest carbohydrates, they break down into small glucose molecules. Plasma in your bloodstream **absorbs** glucose from your gut and carries it to your muscles. Glucose is the only carbohydrate that muscles can use directly for energy.

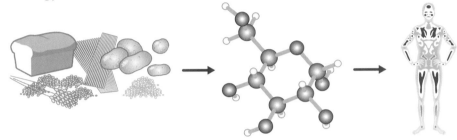

Figure 2 Starch to glucose ($C_6H_{12}O_6$) to energy

Respiration is the process of releasing energy from glucose. Respiration goes on in every cell of your body. Muscles need energy to contract.

Aerobic respiration

Aerobic respiration needs oxygen. 'Aerobic' means 'with oxygen'. This type of respiration happens when you are sitting and walking and when you are jogging. The activities do not need much energy, and there is plenty of time for your red blood cells to carry all the oxygen you need from your lungs to your muscles.

a What kind of respiration needs oxygen?

You can write aerobic respiration as:

glucose + oxygen ⟶ carbon dioxide + water (+ energy)

Respiration symbol equation

You can write the balanced symbol equation as:

$$C_6H_{12}O_6 + 6O_2 \longrightarrow 6CO_2 + 6H_2O \text{ (+ energy)}$$

The aerobic respiration of 1 g of glucose produces 16.1 kJ of energy.

Carbon dioxide and water are waste products of respiration. So our bodies must get rid of them. We store the water in our bladders before we excrete it. We breathe out the carbon dioxide from our lungs.

With training, endurance athletes develop more (and wider) blood capillaries around their alveoli. Their blood can now flow past their lungs more freely, and they can exchange gases at a faster rate.

b What gases do you exchange in your lungs?

Higher

Figure 1 Physiological stamina test. The athlete is breathing through a mouthpiece to assess her oxygen consumption. Electrodes monitor her heart activity and blood pressure.

Anaerobic respiration

Anaerobic respiration does not need oxygen.

This happens when you are running fast, or during any vigorous exercise, when your muscles need more energy than they can get from aerobic respiration. No matter how fast you breathe, or how fast your heart beats, you cannot get oxygen to your muscles quickly enough.

glucose $\longrightarrow$ lactic acid (+ energy)
(Notice that there is no need for oxygen.)

Higher

Lactic acid

Anaerobic respiration takes over from aerobic respiration when you are short of oxygen in your muscles. This lack of oxygen is called an **oxygen debt**. Anaerobic respiration provides less energy than aerobic respiration. Anaerobic respiration only produces 1.2 kJ for every gram of glucose, as chemically it does not fully break down. **Lactic acid** ($C_3H_6O_3$) is produced and builds up in your muscles which makes your muscles ache and gives you cramp. The lactic acid must be broken down using oxygen once you have stopped exercising.

In sport, both aerobic and anaerobic respiration work together to provide energy. However, as your effort increases, aerobic respiration decreases. In a long-distance race you may use 70% aerobic energy.

However, aerobic training increases your ability to get more oxygen to your muscles, which helps in endurance events.

c Why does aerobic training help you to exercise for longer without tiring?

If a person stops training, over a period of time their body reverts to its pre-training state. There will be a loss in muscle mass and a decrease in the ability to deliver oxygen.

d What happens to athletes who are injured for some time?

Summary questions

1 Copy and complete the sentences using the words below:

   glucose   energy   sugar   starch

   .... and .... are carbohydrates. These break down to .... when digested, which respiration converts to ....

2 If more oxygen enters your bloodstream each second, what happens to the oxygen reaching your muscles?

3 What is the difference between aerobic and anaerobic respiration?

4 From what sort of respiration do you think a 400 m runner gets most of their energy?

5 **a** You know that more exercise makes your muscles larger. What happens to the size of your heart as you exercise?
   b A larger heart pumps out more blood every beat. How does this affect the amount of oxygen delivered to your muscles every second?
   c How does aerobic training increase your ability to get oxygen to your muscles?

⚯ **links**

For more information on oxygen debt, see 2.4 Recovery after exercise.

?? Did you know ... ?

A normal resting heart rate is about 60–80 beats per minute. However, rates as low as 50 beats per minute are normal in athletes. Exercise has improved their hearts' ability to pump blood. So now fewer heart beats supply all the oxygen that they require.

Key points

- The heart and lungs provide glucose and oxygen to the muscles.

- Aerobic respiration needs oxygen. Anaerobic respiration does not need oxygen.

- Anaerobic respiration takes over when an oxygen debt occurs. [H]

Recovery after exercise

Learning objectives

- What is oxygen debt? [H]
- How do you control your blood glucose levels?

??? Did you know ... ?

Anaerobic respiration (in which glucose —→ lactic acid (+ energy)) only gives about one-thirteenth as much energy as aerobic respiration. The lactic acid pain is unlucky too. That can really hurt!

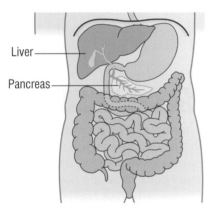

Figure 1 The liver and pancreas

Liver

Pancreas

Figure 2 This girl is injecting herself with insulin. This makes her liver remove the glucose from her blood. Too much insulin could make her dizzy. A sweet can give her a quick glucose boost.

You have seen that anaerobic respiration releases energy from glucose without the use of oxygen. This happens in your muscles during strenuous exercise, when oxygen cannot be supplied quickly enough for aerobic respiration. During anaerobic respiration, the build-up of lactic acid makes your muscles ache.

Oxygen debt

After vigorous exercise you have an 'oxygen debt', because your lungs and heart rate could not keep up with the demand earlier on. You need oxygen to break down lactic acid into harmless carbon dioxide and water.

Before your muscles can operate effectively again, you must:

1 Remove lactic acid. ⎫
2 Replace energy. ⎭ takes about 20 minutes

3 Top-up haemoglobin with oxygen. ⎫
4 Replenish stores of glycogen. (More on this later.) ⎭ takes over 24 hours

Warming down or jogging slowly after you exercise keeps your heart rate and breathing rate up. An active recovery **repays the oxygen debt**, removes lactic acid and replaces your energy more quickly. More lactic acid remains in your blood if you don't warm down actively.

Glycogen

Your body cannot store glucose in your blood. Instead you store it as a carbohydrate, called **glycogen**, in your liver (where 20% is stored) and your muscles (where 80% is stored). Glycogen is converted back into glucose when blood glucose levels fall to a low level. The liver is the largest **organ** in your body.

Two hormones, **insulin** and **glucagon**, control the amount of glucose in your blood. These hormones (or chemical messengers) are made in your pancreas.

Diabetes is a disease where people don't produce enough insulin. Their blood sugar levels can rise to dangerous levels, causing blackouts and blindness.

Controlling blood sugar levels

Your pancreas monitors and controls the glucose concentration in your blood.

Insulin

Digesting carbohydrate foods puts glucose into your blood. When there is too much glucose in your blood, your pancreas secretes the hormone insulin. Insulin makes your liver convert soluble glucose into insoluble glycogen and store it.

Higher

Glucagon

Exercise removes glucose from your blood. When there is too little glucose in your blood, your pancreas secretes the hormone glucagon. Glucagon makes your liver convert glycogen into glucose, which is then released into the blood stream.

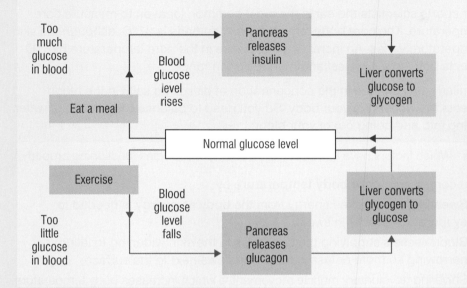

Figure 3 The control of blood sugar levels

a i Which two hormones control the concentration of blood glucose?

ii Where are these hormones made?

Diabetic athletes taking insulin, like Olympic gold-medal rower Sir Steve Redgrave, take special precautions before starting a workout programme. That's because they need to control their blood glucose levels carefully.

b What happens to the concentration of blood glucose when not enough insulin is produced? Why is this dangerous?

Summary questions

1 Which kind of respiration occurs in your body when:
 a you are sitting still
 b you are running fast?

2 What substance makes your muscles ache after hard exercise?

3 What do we mean by an 'oxygen debt'? **[H]**

4 Why does your breathing rate and pulse rate remain high while you recover after exercise? **[H]**

5 What does insulin do when your blood sugar levels are too high?

AQA Examiner's tip

Reference to the hormone glucagon is **Higher Tier only.** Glucagon does the opposite job to insulin.
Both Higher Tier and Foundation Tier candidates need to know how blood glucose levels are controlled by insulin.

Figure 4 Sir Steve Redgrave, five times Olympic gold medallist

∞ links

For information on the dip-stick method for testing blood glucose levels, see 2.12 Standard procedures: for maintaining health and fitness.

Key points

● After vigorous exercise you have an 'oxygen debt' where the lactic acid produced by anaerobic respiration must be broken down by oxygen. **[H]**

● Your body cannot store glucose in your blood. Instead you store it as a starch, called glycogen, 20% in your liver and 80% in your muscles.

● The hormones insulin and glucagon control your blood glucose levels.

Controlling temperature and fluid levels 🇰

Learning objectives

- How do you maintain a constant body temperature?

- How do you maintain the correct amount of water in your body?

❓❓❓ Did you know …?

The thermoregulatory centre in your brain helps to monitor and control your body temperature.

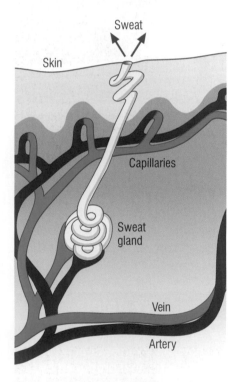

Figure 1 Your skin helps to control your temperature

Your body works best when the conditions inside it (the internal environment) remain the same.

For sports scientists the ear is the most common location to measure core temperature. The core body temperature in humans is 37°C, although the skin temperature varies. An increase or decrease in the core temperature by 1°C affects both your physical and mental performance.

Similarly, any change in the concentration of dissolved salts in the blood affects the working of your body. So you need to balance the amount of water going into and going out of your blood.

a Which factors have to be controlled to keep our bodies functioning properly?

You control your **core body temperature** by:

- Sweating, which takes energy from the body as energy is needed to evaporate the water in sweat.
- Blood vessels supplying the capillaries in the skin widening (dilating) or narrowing so more or less warm blood flows next to the surface.
- Shivering (involuntary muscle movements), which increases body temperature.

You maintain the **level of water** in your body by:

- Gaining water in food and drink.
- Losing water in urine, faeces and sweat.

Activity

Maintaining temperature

If your core body temperature is too hot

When you get out of the shower you feel the effect of **evaporation** causing you to cool down. When you get too hot you sweat a lot! The sweat also takes energy from your skin as the water evaporates.

Feel the saliva in your mouth with your tongue.

Now lick the back of your hand. Does it feel warm or cool?

Where does the water get energy from to evaporate?

When you get too hot the blood vessels supplying your **capillaries** just below your skin expand. More blood flows near the surface. Then the excess energy in your blood **conducts** to the surface where it **radiates** away.

Your face is probably warmer than the table you are sitting at. Hold one hand near your face (but not touching), the other near the table. Which hand feels warmer?

If your core body temperature is too cold

When you get too cold you stop sweating. Think of the colour of your skin when you are cold. When you get too cold the blood vessels supplying your capillaries just below your skin contract. Now very little oxygenated red blood flows near the surface, which limits the energy transfer and your skin will look less pink.

b What two changes in your skin help to keep you cool during exercise?

Competing in a cold climate

Think of a mountaineer on an exposed rock face. If he gets too cold, hypothermia (low temperature) will set in. 'Hypo' means low. Shivering can no longer maintain his body temperature. His skin alone won't keep him warm in the cold. He hasn't got thick insulating hair that traps air like a polar bear!

c How do ski suits keep skiers warm on cold mountain slopes?

Competing in a hot humid climate

Energy transfer by radiation is not possible if the environment is hotter than the athlete. Similarly, the more humid the environment, the less the athlete's sweat evaporates. (Humidity is a measure of the amount of water in the atmosphere.)

If an athlete cannot cool down by radiation or evaporation, their core body temperature will increase by 1 °C every six minutes. Muscle cramp and heat exhaustion are likely.

Fortunately, pouring cold water over you helps. In fact this is 25 times more effective in cooling you down than your body is capable of naturally. A good breeze also helps, as it aids the evaporation of water in sweat. Blow across your hand and see. That's why we use fans in hot weather!

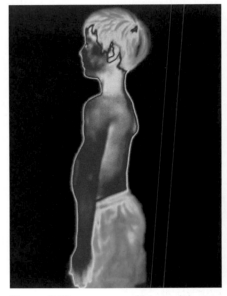

Figure 2 Thermogram of boy emitting infrared radiation after playing sports

Controlling the amount of water in your blood

You control your own water input by the amount you eat and drink. Your **sweat glands** and **kidneys** control your water output. Kidneys are filters that **clean** the blood as it flows past. In fact they filter over 200 litres of blood each day and dump about 2 litres of toxins and excess water into the urine.

You store urine in your **bladder** until you empty it in the toilet. So your body controls the amount of water in your blood by balancing the amounts taken in through food and drink and the amounts given out in urine and sweating.

d Where is urine made?

e What is urine mainly made of?

Summary questions

1 Why do your core body temperature and blood water levels need to remain steady?

2 Muscles put water into your blood during aerobic respiration. In what two ways can this water leave your body?

3 Why does sweating keep you cool?

4 Imagine it is a hot day. You sweat a lot playing a sport. Why is your urine a darker colour when you go to the toilet?

Key points

- During exercise your core body temperature increases. You sweat and blood vessels supplying the capillaries below your skin expand (dilate). Energy transfer by evaporation and by radiation cause cooling.

- You remove water in your body in sweat and in urine.

2.6 Physiotherapy

Learning objectives

- What common injuries do people suffer from?
- How do physiotherapists treat injured athletes?
- How are bones supported?
- Why is the skeleton important?

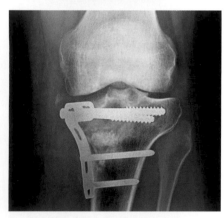

Figure 1 A pinned broken knee

Injuries to the human body

Physiotherapists regularly treat people suffering from muscle, ligament or joint injury. The physiotherapist may massage their muscles or manipulate their joints. This helps to relieve muscle pain and stiffness and encourages blood flow to injured parts of the body to aid recovery.

Sport and exercise can sometimes cause injuries. Sports injuries can happen as a result of:

- not warming up properly before exercising more vigorously
- pushing too hard for your current level of fitness, or
- using inadequate equipment.

Serious injuries occur as a result of a sudden impact or awkward movement. Persistent injuries develop over time, often due to continual use of the same joints or muscle groups.

Sports injuries include damaged ligaments, pulled or torn muscles, torn cartilage, ruptured tendons, dislocated joints and fractured bones. Serious injury may require surgery – using metal plates or pins to support healing bones and Kevlar fibres to replace damaged ligaments.

   a Why do sports injuries occur?

   b What kinds of sports injuries can athletes suffer from?

   c Why are pins sometimes fitted to broken bones?

   d Why does rehabilitation take time?

Scientists @ work

Rehabilitation

If you break a bone you are not 'fighting fit' as soon as your plaster cast is removed! You need light exercise during rehabilitation to build up your muscles that have shrunk through lack of use.

Injured athletes take time to heal physically and emotionally. Many professional footballers have taken months to recover their movement and regain their abilities. They cannot compete until they return to full health, otherwise they may make any damage worse. Sports physiotherapists often supervise their exercise during recovery.

The skeleton

Your skeleton supports your body and protects your vital organs, like your brain, heart and lungs. Muscles allow your body to move by shifting the position of bones in your skeleton. The skeleton accounts for about 14% of an adult's body mass, and half of this mass is water.

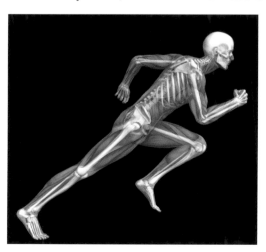

Figure 2 Your skeleton supports your body and protects your vital organs. Your muscles allow your body to move.

   e What purpose does your skeleton serve?

Joints

Joints allow your limbs to move in different directions. The strong fibres that hold your bones together at a joint are **ligaments**. Ligaments keep joints stable. The strong fibres that attach muscles to your bones and make them move are **tendons** (see Figure 3).

The synovial joint

Your knee is a synovial joint, as are many others, like your shoulder, elbow, wrist, hip, and even your knuckles. They contain a slippery layer, called **cartilage**, which stops your bones rubbing together and lets them move freely over each other. Cartilage reduces friction. **Synovial fluid** lubricates the joints and also helps reduce friction between cartilage at the joints. The synovial membrane, that lines all the non-cartilage surfaces at these joints, produces the synovial fluid.

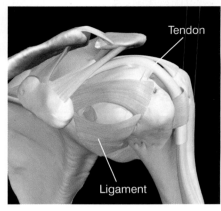

Figure 3 The ligaments and tendons of the shoulder joint

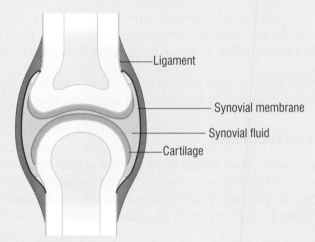

Figure 4 A synovial joint, where bones meet

f How does cartilage and synovial fluid protect joints?

Did you know ... ?

'The hip bone's connected to the thigh bone'... but did you know, there are no muscles in fingers. The muscles in the palm of your hand connect to the bones in your fingers by tendons, which pull on your fingers to move them.

Summary questions

1 A sprained ankle occurs when ligaments suddenly over-stretch. State two other examples of sports injury.

2 Why does your body have joints?

3 Look at Figure 5 below:
   a Name a hinge joint.
   b Name one type of ball and socket joint.

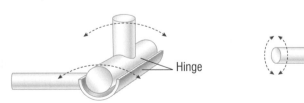

Figure 5 A hinge joint and a ball and socket joint

4 What is the difference between a tendon and a ligament?

5 **a** How is friction at synovial joints, like the knee, reduced?
   b How does less friction benefit synovial joints? [H]

Key points

- Common injuries include damaged ligaments, pulled or torn muscles, torn cartilage, ruptured tendons, dislocated joints and fractured bones.

- Physiotherapists massage muscles and manipulate joints when treating injured athletes.

- The skeleton supports the body and protects vital organs.

- Tendons attach muscle to bone and make them move. Ligaments link bones to stabilise joints.

- Cartilage and synovial fluid prevent friction between bones at synovial joints such as the knee. [H]

2.7

Biomechanics – the science of human movement

Figure 1 Using a dumb-bell

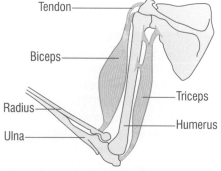

Figure 2 A good example of an antagonistic pair is the biceps and triceps. One makes the elbow bend, the other makes it straighten. This raises and lowers your forearm.

⚙⚙ **links**

For more information on the grip test method to measure the strength of a muscle see 2.12 Standard procedures for maintaining health and fitness.

Muscles and how your body moves

Humans have three different kinds of muscle:

- **Cardiac muscle** – cause an adult's heart to beat about 70 times every minute and pump five litres of blood per minute.
- **Smooth (or involuntary) muscle** – work on their own, to push food through your gut and expel urine from your bladder.
- **Skeletal (or voluntary) muscle** – when you decide to move, skeletal muscles pull on your bones. These muscles work quickly and strongly, but they soon tire.

A muscle can only pull on a bone, it cannot push. For this reason muscles always work in pairs. As one muscle contracts the other relaxes. Pairs of muscles are called **antagonistic pairs**, because they always oppose each other.

    a Lift your forearm. Feel your biceps contract and your triceps relax. What happens to your biceps and triceps when you extend your forearm?

Levers and moments

Your forearm works like a simple **lever**, pivoted at your elbow and forced up by your biceps. The force (or effort) exerted by the biceps is more than the weight (or load) that is being moved. This is because the effort is applied near to the pivot. We can explain this by calculating the **moments** for both the effort and load.

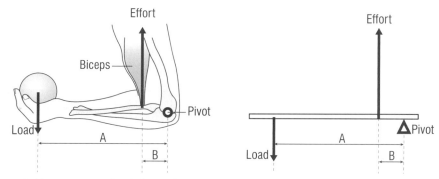

Figure 3 Forearm as a simple lever. A and B are the distances from the pivot to where the forces act.

All levers turn or twist about a pivot. The turning effect of a force is called its **moment**.

The moment is calculated using the equation:

    Moment = force × distance
      (Nm)   **(N)**   **(m)**   This distance is the perpendicular (or 90°) distance to the pivot.

We measure turning effects or moments in newton metres (Nm).

If two moments match each other, they balance a lever so it does not rotate. This happens with see-saws and when limbs are held steady. If the two moments do not match each other then there will be movement around the pivot. This happens when we move our arms, legs and numerous other joints.

Your skull pivots on the top of your spine. If you relax the muscle in the back of your neck then your head tips forward. This is because the centre of mass of your head is in front of the pivot and its weight creates a moment. To keep your head upright the muscle in the back of your neck must pull the back of your skull down.

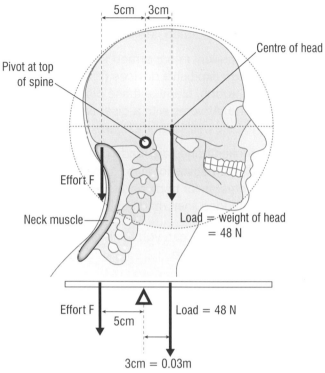

Figure 4 Moments acting on an adult head

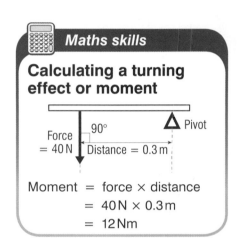

AQA Examiner's tip

You don't need to know the names of all 206 bones in your body. Once you know the functions of the skeleton, the structure of a joint and how muscles work, you can link that knowledge to an example the examiner uses.

Maths skills

Calculating a turning effect or moment

Moment = force × distance
= 40 N × 0.3 m
= 12 Nm

b Suppose your head weighs 48 N (equivalent to 4.8 kg) and this force acts 3 cm from where it pivots. Show that its turning effect or moment is 1.44 Nm.

c To support your head the muscle in the back of your neck must also exert a moment of 1.44 Nm. The muscle pulls on your skull 5 cm behind where the skull pivots. With what effort, F, does this muscle pull down?

Summary questions

1 Why are at least two muscles needed at a joint?

2 A force is applied to the lever as shown in the diagram. Calculate the moment in newton metres.

3 Calculate the turning effect (moment) of a force of 30 N that acts at a perpendicular distance of 0.2 m from a pivot.

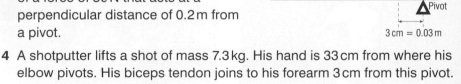

4 A shotputter lifts a shot of mass 7.3 kg. His hand is 33 cm from where his elbow pivots. His biceps tendon joins to his forearm 3 cm from this pivot. (Assume 1 kg weighs 10 N.)
  a Calculate the moment that the shot exerts.
  b Calculate the force applied by his biceps when he holds the shot still.

Key points

● Skeletal muscles work as antagonistic pairs. They pull on bones to make them pivot.

● A moment is the turning effect of a force. Calculate it using:

Moment = force × perpendicular distance to the pivot

● Moments have units of newton metre or Nm.

2.8 Prosthetics

Learning objectives

- What are the properties of materials for artificial joints?
- What materials are artificial joints made from?
- What ethical issues affect competitive sports?

Did you know ...?

In the word paralympics, 'para' stands for 'parallel or equal to', not 'paralysed'. This distinction is important to Paralympic sports stars. The Paralympic Games were first held in 1948 to coincide with the first London Olympics. Nurses were employed as referees and doctors handed out the medals.

Prosthetics

One person in every thousand will have a major injury at work this year. One in forty of these will have a limb amputated. Most of these will be young workers.

This type of injury is all too frequent in our armed forces. Already this century roadside bombs or improvised explosive devices (IEDs) have seriously injured thousands of British troops. Many of these disabled young men and women will take part in sports. Athletes with a disability use specially adapted equipment to allow them to take part in their chosen sport.

Small pins and rods to fix broken bones are made of steel, while aluminium and titanium are lightweight alternatives for larger prostheses. 'Prostheses' are artificial (man-made) replacements for body parts. Below-knee amputees need a comfortable 'soft-socket' padded liner, made of Pelite, a poly(ethylene) foam.

a How many types of prostheses can you think of? Here are two to start you off: an artificial hip, false teeth.

Artificial joints

Artificial joints need to be lightweight, yet strong. Acrylic, epoxy and polyester are plastic **polymers**. They can be moulded to fit damaged knee joints and are reinforced with carbon, glass or Kevlar fibres.

Nowadays many elderly people benefit from replacement hip and knee surgery. The most common cause of joint damage in the elderly is osteoarthritis. This is when the bones lose their protective cartilage, are damaged, and become inflamed causing severe pain and loss of mobility. Then only joint replacement surgery can improve their quality of life.

The surfaces of the artificial joint are shaped to give the person the same movement as a natural joint. Modern materials like zirconium alloy help. Zirconium is an element, similar to titanium, which absorbs oxygen at its surface to create a smooth ceramic surface.

b Describe a possible remedy for patients with worn or injured joints.

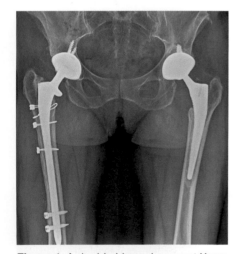

Figure 1 A double hip replacement X-ray of a 63-year-old patient

Figure 2 Titanium alloy composites make this highly manoeuvrable wheelchair both lightweight yet strong

Figure 3 Double amputee Oscar Pistorius, the 'Blade Runner', is 'the fastest man on no legs'

A student's water-skiing accident revolutionised prosthetic foot design. The 21-year-old, who lost his leg, teamed up with an aerospace engineer to design a lightweight, strong, yet flexible L-shaped foot. Their 'Flex-Foot', made of carbon fibre composite, was inspired by the shape of a cheetah's rear leg.

Nowadays, over 90% of amputee athletes worldwide wear a Flex-Foot in competition. When an athlete puts his or her weight down, energy transfers, which literally puts a spring in their step. This also stops the athlete's body jarring, and prevents repetitive strain injury in muscles and tendons.

c What features of the Flex-Foot design help when sprinting?

d What could happen if the Flex-Foot was stiff as opposed to flexible?

Investigation

Flex-Foot

Bend a paper clip into a Flex-Foot shape. Push down slightly and feel the spring. Push down more and notice a durability problem. Explain this problem.

Activity

The Flex-Foot debate

Does Flex-Foot give Oscar Pistorius an unfair advantage over able-bodied runners? The **International Association of Athletics Federations** (IAAF) thought it did.

In 2007 they introduced a rule banning 'any technical device that incorporates springs, wheels or any other element that provides a user with an advantage over another athlete not using such a device'.

● Do you agree with the IAAF's ruling that athletes should always compete 'on a level playing-field'? Justify your answer.

● Is it unethical to use advanced materials for artificial limbs and joints when competing in sports? Consider other sports, and factors like cost, risk of injury, etc. in your discussion.

 Did you know ...?

Until 1967, it wasn't illegal for athletes to use performance enhancing drugs. The 2009 World Anti-Doping Code regards doping as not only unethical, but dangerous to health.

Activity

Ethics – evaluating the issues

When does performance enhancing behaviour become cheating? Most people agree that using 'designer' drugs, like steroids, to improve performance is unethical. Nobody really wants 'engineered' sports stars. Yet, surprisingly, laser eye treatment is not banned in archery and shooting, nor are inhalers banned for asthmatics.

● Where should you draw the line to prevent one athlete gaining an unfair advantage over another?

Key points

● Apart from being comfortable, artificial joints must be lightweight and strong.

● Plastic polymers can be moulded to joints. Kevlar fibres add reinforcement. Titanium is the best light, yet strong, metal. Poly(ethylene) foam makes a comfortable soft-socket pad for amputees.

● Ethical issues arise when athletes wearing prosthetic limbs made from advanced materials want to compete alongside able-bodied athletes.

Summary questions

1 What is a prosthesis?

2 Where might a plastic polymer, like acrylic, be used in prosthetics?

3 Flex-Foot is an artificial sports limb. What properties does it have?

4 Why do sports wheelchairs need to be light and strong?

2.9 Nutrition for exercise and fitness

Learning objectives

- What are nutrients?
- What is a balanced diet?
- How do athletes record their food intake?

Scientists @ work

Dieticians

Dieticians advise us to eat a nutritious balanced diet in order to stay healthy. What does this mean, and why is it important?

Figure 1 High protein breakfast of a powerful sprint athlete or a weightlifter

Figure 2 Bread, pasta and sugar are all sources of carbohydrates

∞ links

For more information on how food analysts detect the presence of nutrients in your food, see 4.16 Standard procedures used in food science.

Balanced diet

Your body requires a variety of **nutrients** in order to carry out the vital life processes such as respiration (the release of energy), movement, growth and repair of body tissue. These nutrients are divided into five groups:

- Carbohydrates, from sugar and starch, provide your main source of energy. Everything from respiration to physical activity requires energy. Strenuous exercise uses up to 4 g of glucose every minute. If your body contains more glucose than it needs, it is converted into glycogen. This energy reserve is stored in your liver and muscles. Glycogen is the main source of energy used by muscles. If you train with low glycogen stores you will feel tired. Your performance will be at a lower level and you will be more prone to injury and illness.

a Why do more strenuous activities require more energy?

b Why is it necessary for athletes to take in enough energy every day?

- **Proteins** aid the growth and repair of body tissues, and provide some energy. You get protein in meat, fish, eggs, milk, cheese, cereal, peas and beans. The recommended daily protein requirement for people aged 15 and over is 0.8 g per kg of body weight. Your body uses the protein you eat to make specialised protein molecules such as hormones and enzymes. These molecules carry out specific jobs in your body. For example, the hormone insulin is a protein which regulates the level of glucose in your blood.

- **Saturated and unsaturated fats** provide you with a store of energy. They also help to keep you warm by providing a layer of insulation under your skin. Fat covers your vital organs, like your kidneys and heart. This protects them from damage. No more than 30% of your energy intake should come from fat.

- **Vitamins** and **minerals** protect vital organs by keeping them working efficiently.

- **Water** helps to remove toxins from the blood. Insufficient water in athletes causes heat exhaustion, signalled by cramp, nausea and headache. Squash, fizzy drinks and fruit juices provide water. However, about one-third of your water intake comes from fruit and vegetables. If your urine is dark yellow, you need more fluid in your diet.

Fibre

Fibre, from plant cell walls, is an essential component of your diet. (It is *not* classed as a nutrient as it does not provide us with any goodness.) Fibre adds bulk to your food so that waste can be pushed out of the digestive system more easily. Fibre also absorbs poisonous waste and makes you feel full, which helps to stop you overeating.

Nutritional tips

1 Don't skip meals. It doesn't help to control your weight.
2 Most of us don't drink enough water. Aim to have six to eight glasses a day.

3 'Eat up your greens!' Fruit and vegetables keep you healthy. Aim for five portions each day.
4 'Live bio' yoghurt puts good bacteria in your gut to aid digestion.
5 Buy 'low fat' dairy products to limit fat intake.
6 'A *little* of what you fancy does you good', but alcohol is not a fuel for exercise.

Dietary intake

Sports nutritionists and dieticians study the nutrient intake of athletes. Their advice can help an athlete to maximise his or her performance. The simplest way to monitor your diet is to keep a diary.

You can use a **diet diary** to track how many kilocalories (kcal) or kilojoules (kJ) of energy you take in from each source of food (1 kcal = 4.2 kJ).

For some foods you will need to look up the information in a food table which shows details of the energy (in kcal), protein (in g), carbohydrate (in g) and fat (in g) for 100 g of the food.

However, today most foods have kcal and kJ written on their wrappers.

Activity

24-hour diary

Copy the table and record everything you eat and drink in the next 24 hours.

Food	Quantity	Total energy (kcal)	Carbohydrate (g)	Fat (g)	Protein (g)

Weigh out quantities of food like breakfast cereal first. Note the information on the nutrition labels. Recall any foods you didn't have time to note down at the time. Then use a food table (search online), or food label to find the data you need.

- Why do you think you shouldn't leave more than 24 hours between eating and recording the diary information?
- Weigh yourself in kilograms. Multiply your mass in kilograms by 0.8. This is your daily **target** protein intake in grams. Total your protein column in grams. Have you eaten more or less protein than your target?
- Total your carbohydrate and fat columns. What percentage is fat?

Summary questions

1 What happens to carbohydrates which are not required by the body as an immediate energy source?

2 a Look at the nutrition label in Figure 3. How much energy do two sheets of lasagne provide?
b Look at the 'energy value' line in kcal. What is the average daily energy needed for a woman? Include the units.
c A sheet of lasagne has a mass of 16 g. What percentage of protein is in a sheet of lasagne?

3 The energy content of ice cream is 170 kcal per 100 g. Suppose you only eat 1 g (unlikely), its energy is 1.7 kcal. 25 g is more likely. How much energy will 25 g of ice cream give you (in kcal and kJ)?

AQA Examiner's tip

Questions relating to balanced diets are very common. Make sure you can name the main food groups, their role in the body and examples of foods within each group.

NUTRITION

TYPICAL VALUES	PER SHEET (APPROX 16 g)	PER 100 g (UNCOOKED)
Energy Value (calories)	230 kJ (55 kcal)	1470 kJ (345 kcal)
Protein	2 g	12 g MEDIUM
Carbohydrate (of which sugars	12 g 0.3 g	72 g HIGH 2 g) LOW
Fat (of which saturates	0.2 g Trace g	1 g LOW 0.3 g) LOW
Fibre	0.5 g	3 g MEDIUM
Sodium	Trace g	Trace g LOW

GUIDELINE DAILY AMOUNTS

	Women	Men	Approx. Per sheet
Calories	2000	2500	55
Fat	70 g	95 g	0.2 g
Salt	5 g	7 g	Nil

These figures are for average adults of normal weight. Your own requirements will vary with age, size and activity level.

Figure 3 Nutrition label on a packet of lasagne

Key points

- Nutrients are needed to carry out the vital life functions: respiration, movement, and the growth and repair of body tissue.
- A balanced diet contains carbohydrates, fats, proteins, vitamins, minerals and water.
- Only with a correct intake of nutrients can an athlete reach their full potential.
- The simplest way to monitor your diet is to keep a diet diary.

2.10 Energy requirements

Learning objectives

- How do people's energy requirements depend on their weight and their level of exercise?

- Why do athletes increase their intake of complex carbohydrates before competing?

Maths skills

Calculating BER

BER for 1 kg of body mass is **5.4 kJ per hour.**

- Multiply 5.4 by 24 for BER per day.
- Multiply 5.4 by the mass in kg to give BER per hour.

Did you know … ?

Sports scientists know the energy cost of many sports activities by monitoring the oxygen consumption of athletes. For example, to run 1 km you breathe in an extra 15.6 litres of oxygen (per kg approximately). For every extra 7 kg of body mass, you expend 10% extra energy exercising.

Every physical activity requires energy. The amount depends on the duration and type of your activity, as well as your body mass.

When resting, your **basic energy requirement (BER)** for every kilogram of body mass is 5.4 kJ per hour. So your BER per day (per kilogram) is:

5.4 × 24 hours   =   130 kJ every day

So the BER of a girl of mass 50 kg would be:

50 × 5.4 = 270 kJ per hour and 6480 kJ per day

> **a** What is the BER for a man of mass 65 kg in 24 hours? Show that this is about 8450 kJ/day.

The extra amount of energy you require when exercising depends on the type of activity and the energy that you expend. To run 1 km you need an extra amount of energy = (your mass in kg) × 330 kJ.

Your body needs more energy (and more carbohydrate) than a basic amount if you train or work hard physically. Your basic need is your BER, so:

Your personal energy requirement   = BER + an extra amount

For every kilogram of body mass you need a BER of 130 kJ every day.

For every hour you exercise you need an extra 36 kJ for each kilogram of body mass.

Suppose you weigh 50 kg and exercise for two hours today:

Your basic need BER = 50 × 130 = 6500 kJ

Your extra amount = 2 × 36 × 50 = 3600 kJ (i.e. hours × 36 × kg)

Your personal energy requirement = 6500 + 3600 = 10 100 kJ

> **b** Why does an athlete require a higher carbohydrate diet than the average person?

Practical

Personal energy requirement

- Weigh yourself. What is your body mass in kilograms?
- What is your **basic** daily energy requirement?
- How many hours of exercise did you do yesterday?
- What **extra** energy did you need?
- What was your **personal energy requirement** yesterday?

Body mass index (BMI)

Being overweight increases your risk of heart disease and other illnesses. Your **body mass index (BMI)** is a good indicator of your total body fat. Body mass index reveals your ideal weight. Your BMI takes into account both your mass (m in kg) and your height (h in m):

$$\text{Body mass index (BMI)} = \frac{\text{mass (kg)}}{(\text{height in m})^2} = \frac{m}{h^2} = \frac{m}{h \times h} \quad (\text{units kg/m}^2)$$

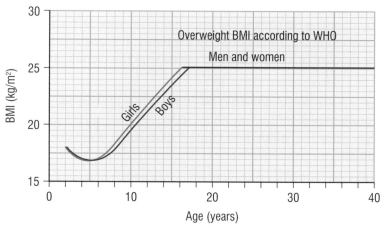

Figure 1 Graph showing BMI information from WHO (World Health Organisation)

There are limitations of using BMI. For example, a fit weightlifter is likely to be short, muscular and heavy. As muscle weighs more than fat, his ideal BMI will be over 25.

Being well underweight or overweight creates problems, so if your BMI is less than 17.5 or more than 25, that's not good. But you knew that already, above all, don't worry. It is unhealthy to have anxiety over a less-than-perfect body image. A healthy mental attitude is just as important as physical fitness.

Maths skills

BMI

If your weight is 80 kg and your height is 1.8 m,

your BMI $= \dfrac{80 \, \text{kg}}{(1.8 \, \text{m})^2} = \dfrac{80}{3.24} = 24.7 \, \text{kg/m}^2$

This is close to the overweight limit, so think about your diet and exercise.

Summary questions

1 What do the terms **a** BER **b** BMI stand for?

2 In which sports events do athletes need large amounts of carbohydrate in their diet?

3 Imagine it is the morning of a sports competition. You do not feel like eating.
   a Why might you feel like this?
   b Why should you eat anyway?

AQA *Examiner's tip*

Remember to calculate Basic Energy Requirement using:

BER = kg × 5.4 kJ × hours

Key points

● All athletes need a carbohydrate-rich diet that provides enough energy. Your daily energy requirement depends on your body mass and level of exercise.

● BER (Basic Energy Requirement) is 5.4 kJ/hour for every kilogram of body mass.

● BMI (Body Mass Index) is (mass in kg) ÷ (height in m)² (units kg/m²).

2.11 Sports drinks and sports diets

k

Learning objectives

- What do isotonic sports drinks contain?

- How does a normally balanced diet compare with the diets of different athletes?

- Why do athletes increase their intake of complex carbohydrates before competing?

- Why do some athletes eat a high-protein diet?

??? Did you know ... ?

A 2% loss of body fluid saps performance. A 5% loss causes heat exhaustion.

Figure 1 Isotonic sports drinks

Sports drinks

Imagine you are dehydrated after sweating. Water is not the best drink to replace your lost body fluid. It quenches your thirst before you have drunk enough. Water also encourages your kidneys to produce urine, which delays re-hydration.

Your body contains 50–75% water depending on your age and body fat. The average man needs 2.9 litres of water each day; the average woman needs 2.2 litres. The amount of water you need varies with the climate and your level of physical activity.

Nowadays sports drinks are a big business. They are specially formulated carbohydrate drinks containing water, glucose, electrolytes (certain ions) and flavourings. They re-hydrate and contain glucose to give your muscles a quick energy boost.

Hypertonic drinks (hyper means *high*) contain high levels of glucose to top up athletes' carbohydrate intake. **Hypotonic** drinks (hypo means *low*) contain little glucose, but quickly replace fluids lost by sweating.

Five factors are important in designing a sports drink:

- **Taste** – Fruit flavours make the drink appetising.

- **Glucose content** – This provides instant fuel for energy. Jockeys and gymnasts who sweat, but want to maintain a low body weight, prefer hypotonic drinks with less glucose.

- **Stomach emptying** – Hypertonic drinks with more than 7 g of glucose in every 100 g of water, empty from your stomach slower than water alone. However, the more you drink the faster it leaves your stomach and enters your intestine.

- **Fluid absorption** – Glucose and ions stimulate absorption of fluid into the bloodstream. This speeds up re-hydration.

- **Urine production** – Sodium and potassium salts in the drink reduce urine output. Any solution containing dissolved ions such as Na^+ and K^+ is called an **electrolyte**. Electrolytes conduct electricity. Sports drinks are often called 'isotonic' or 'isoelectronic'. Iso means *same*. Having the same proportion of ions in the drink and in the blood is the ideal.

Practical

Make your own isotonic drink

Use the information above and a combination of water, glucose and electrolytes to make your own isotonic drink. Ensure the equipment you use is thoroughly sterilised. Do this in the food technology department.

a What are the three main ingredients of isotonic sports drinks?

b Why should your drink contain less than 8% glucose?

c For flavouring, why should you use reduced-sugar squash?

Sports diets

All athletes need a carbohydrate-rich diet that provides enough energy.

Carbohydrate/glycogen loading

Carbohydrates break down into glucose, which your muscles use for energy. You store glucose as glycogen in your liver and muscles. Your body only stores enough glycogen for 90 minutes of exercise. Athletes who suddenly tire and 'hit the wall' have depleted their glycogen stores. They must consume carbohydrates during a marathon to maintain their blood glucose levels.

A hypertonic sports drink replaces both the carbohydrate used for respiration and the water lost as sweat. However, a 'carbohydrate/glycogen loaded' diet boosts the athlete's levels of glycogen and more than compensates for any loss during their race.

Figure 2 Paula Radcliffe shattering her previous world-best in the London Marathon

Time	Diet	Exercise
5–7 days before the event	Low-carbohydrate diet	Strenuous exercise
2–4 days before the event	Increase carbohydrates	Reduce exercise
Day before the event	High-carbohydrate diet and 'pasta party' the night before	Complete rest day

d Why do athletes increase carbohydrate intake and gradually reduce their training before competition?

High-protein diets

The longer and the more intensely an athlete trains, the more their protein breaks down. Unless carbohydrate intake is sufficient to meet their energy needs, protein will get used for energy rather than for growth and repair.

However, eating more protein will not increase the size of your muscles! Muscles develop from exercise when there is enough protein in your diet. Power athletes, such as sprinters, train to gain muscle mass. Their typical daily diet might include having: a **breakfast** of bacon and eggs; a **lunch** of salad and double cheeseburger (without bread); and a **dinner** of steak (or fried chicken or fish), with a salad topped in cheese dressing.

A high-protein low-carbohydrate diet puts stress on the kidneys, and can cause headaches, tiredness, dizziness and constipation.

Figure 3 100m hero Usain 'Lightning' Bolt 'powered by chicken nuggets and yams'

e Protein, when combined with exercise, helps to grow muscle. Which types of sportsmen and women need a high-protein diet?

Summary questions

1 Why do athletes sweat during exercise?

2 Why do marathon runners need isotonic drinks during a race?

3 Bananas are easy to eat and rich in carbohydrate. Eat a banana, with plenty of water, one hour before a competition. Explain the benefits of this combination for an athlete.

4 Why does a high-protein diet encourage the over-eating of saturated fat and cholesterol, increasing the risk of heart disease.

Key points

- Isotonic sports drinks contain water, glucose and electrolytes. Hypertonic – high glucose. Hypotonic – low glucose.

- All athletes need a carbohydrate-rich diet that provides enough energy.

- Power athletes, such as sprinters, train to gain muscle mass. They have a high-protein diet.

2.12

Standard procedures for maintaining health and fitness

Healthcare scientists must follow **standard procedures** 'to the letter', otherwise data they obtain from their tests will not be repeatable and reproducible. Sports scientists and fitness coaches often conduct pre-test and post-test baseline measurements of an athlete before exercise and after exercise as he or she recovers. This helps them to assess over time the benefits of the athlete's personal exercise plan.

You have practised applying some of these standard procedures yourself. Below are some more health and fitness standard procedures.

Testing blood glucose levels using urine

The chemical company Bayers makes a **glucose-testing strip** (or dip-stick), called 'Diastix'. It changes colour based on glucose concentration. This dip-stick is used to test glucose concentrations in urine.

Practical

Directions for using Diastix

1 Dip the chemical end of the stick into artificial urine and remove immediately.
2 Remove excess *urine* and wait exactly 30 seconds before comparing with the colour chart.

If you have excessive amounts of glucose in your blood, your kidneys filter some of this glucose into your urine. The concentration of glucose in your urine is not what is actually in your blood.

a Which organ in your body removes excess glucose from your blood?

Measuring muscle strength

Strength is the ability of a muscle to exert a force for a short time.

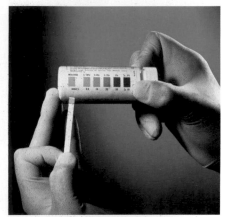

- membrane lets glucose through
- active chemical in paper pad
- plastic backing of the dip-stick

Figure 1 Diastix glucose-testing strip

??? Did you know ... ?

In days gone by, doctors used to test for diabetes by seeing if ants were attracted to the high blood glucose levels in the patient's urine. The other way was to taste it and see if it was sweet!

Practical

The 'handgrip strength test'

The 'handgrip strength test' involves squeezing a handgrip dynamometer as hard as possible:

1 Hold the dynamometer in line with your forearm and hanging by your thigh. Squeeze without swinging your arm.
2 Repeat after one minute. Then repeat with your other hand.

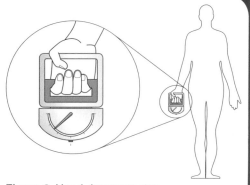

Figure 2 Hand dynamometer

Remembering standard procedures

Use the following table to help you to revise the standard procedures. Only do baseline tests requiring physical activity if you are fit and healthy, and remember to carry out a risk assessment first.

Students need to know:	Topic	Further guidance
How to measure heart rate (pulse)	2.1	Use a pulse meter or 'find a pulse' and count for 30 seconds
How to measure breathing rate	2.2	Count the breaths in one minute. When fitness testing measure before and after exercise
How to measure: a vital capacity (VC) and b tidal volume (TV) of the lungs using a spirometer	2.2	Wear a nose-clip and disinfect the mouthpiece after use. a Tidal volume (TV) is the volume of air in ml (or cm^3) you breathe in and out normally b Vital capacity (VC) is the maximum volume of air you can force out after breathing in as hard as you can
How to measure temperature	2.5	Ear thermometers measure eardrum temperature using infrared sensors. Do not push it down the ear canal
How to measure a moment (or turning effect of a force)	2.7	Moment (in Nm) = force (in N) × distance (in m) The distance is the perpendicular distance to the pivot
How to calculate basic energy requirement (BER)	2.10	Your basic energy requirement BER (for every kilogram of body mass) is 5.4 kJ per hour
How to calculate body mass index (BMI)	2.10	$$BMI = \frac{mass\ (kg)}{(height\ in\ m)^2}$$ $$= \frac{m}{h^2} = \frac{m}{h \times h}\ (units\ kg/m^2)$$
How to measure the glucose content of blood and urine using a dip-stick method	2.12	Dip the Diastix into the urine sample, remove and wait 30 s. Check the concentration in mg/100 cm^3 of urine
How to measure the strength of muscle by the grip test method	2.12	Hold the handgrip dynamometer in line with your forearm and hanging by your thigh. Squeeze it quickly without swinging your arm

∞ links

For information on examples of health and fitness standard procedures look back at 2.1 Your heart (pulse), 2.2 Your lungs (breathing rate, TV and VC), 2.7 Biomechanics – the science of human movement (measuring a moment) and 2.10 Energy requirements (BER and BMI).

Maths skills

See the equations on this spread involving moments and body mass index. Check the topic spreads for details of the calculations.

Remember also how to do calculations involving Basic Energy Requirement BER on Topic 2.10.

BER = 5.4 kJ per hour (for every kilogram of body mass)

= 5.4 × 24 hours

= 130 kJ every day (for each kilogram)

Summary questions

1 Why do scientists follow standard procedures so precisely?

2 a Briefly describe how you measure blood glucose levels using urine.
   b What is more significant to an athlete, the glucose in their blood or their urine?
   c Why do urine glucose levels always lag behind blood glucose levels?
   d Why do urine test strips not show if blood glucose levels are too low?

3 What are the following used for?
   a A dynamometer
   b A spirometer

4 Why are baseline measurements taken pre-test and post-test (before and after an athlete exercises)?

Key points

● A standard procedure is like a recipe – an agreed way of doing something.

● You measure glucose levels of blood and urine using a dip-stick method.

● Use a handgrip dynamometer to measure muscle strength.

● Revise all the techniques listed in this topic spread.

Summary questions

1 Copy and complete the sentences using the words below:

diaphragm   glucose   heart   increase
intercostal   oxygen   pulse

The cardiovascular system refers to the .... and its blood vessels. This, together with the lungs, helps .... and .... to get to our muscles. Our .... and breathing rate .... during exercise. The .... and .... muscles in our thorax allow our lungs to ventilate.

2 Copy and complete the sentences using the words below:

anaerobic   blood   capillaries   glucose   insulin
liver   oxygen   pancreas   respiration   sweat

Respiration may be aerobic or .... When there is not enough oxygen during exercise, an .... debt in the muscles causes anaerobic .... to take over. To maintain a constant body temperature during exercise we .... and .... in our skin open up letting more .... flow near the surface. We control our blood glucose levels with the help of the hormone .... If our blood glucose levels rise, the .... releases insulin, causing the .... to convert .... into glycogen.

3 What is the balanced symbol equation for aerobic respiration? **[H]**

4 What is the effect of 'oxygen debt'? **[H]**

5 How does the hormone glucagon control blood glucose levels? **[H]**

6 Copy and complete the sentences using the words below:

biceps   bone   force   pivot   relaxes
tendons   triceps   turning

Our .... and .... are examples of antagonistic muscles. They work in pairs. When one .... the other contracts. Sports injuries include damaged ligaments, which join bone to .... and ruptured .... which attach bone to muscle. When a tendon pulls on a bone a .... effect occurs called a moment. This is the .... multiplied by the distance to the ....

7 Describe the structure and function of a synovial joint. **[H]**

8 Calculate the moment (in Nm) when a biceps tendon applies a force of 500 N to a bone in the forearm. The tendon joins the bone 3 cm from where the arm pivots at the elbow.

9 What is the daily basic energy requirement (BER) for a lady of mass 55 kg (BER = 5.4 kJ per hour for every kilogram of body mass)?

10 a Calculate the body mass index (BMI) for an adult male of mass 85 kg and height 190 cm.
   b Is he overweight?

Key practical questions

1 Describe how to take baseline measurements of:
   a heart rate (pulse)
   b breathing rate
   c tidal volume (TV) and vital capacity (VC) using a spirometer
   d temperature with an ear thermometer
   e body mass index (BMI)
   f glucose content using a dip-stick
   g muscle strength using the handgrip test.

AQA Examination-style questions

1 A sports physiologist measures the changes to an athlete's heart rate during exercise.

  a The athlete has his heart rate measured before he does any exercise.

    i Describe how the sports physiologist would use the athlete's pulse to calculate heart rate.   (2)

    ii Why can the pulse only be felt in an artery?   (1)

  b The sports physiologist notices that the heart rate increases during exercise. Explain this change.   (2)

  c The sports physiologist notices that the heart rate does not immediately return to normal when the athlete stops exercising. Explain why. **[H]**   (3)

2 Sports nutritionists advise athletes on how their diet can be improved to maximise performance.

  a An athlete weighs 90 kg. Calculate his Basic Energy Requirement for one day.   (2)

  b Describe how the athlete would have been able to provide the nutritionist with the information needed to analyse his diet.   (2)

3 Sports physiotherapists treat skeletal–muscular injuries.

  a Give **two** functions of the skeleton.   (2)

  b Describe how muscles work together to bend the leg at the knee.   (1)

  c Footballers often wear shin pads to protect themselves from injuries. There are different varieties of shin pad. Describe how to test and compare two types of shin pad to see which would be more effective at protecting against injury.   (3)

4 An athlete exercised on a running machine. The machine increased in speed every 60 seconds.

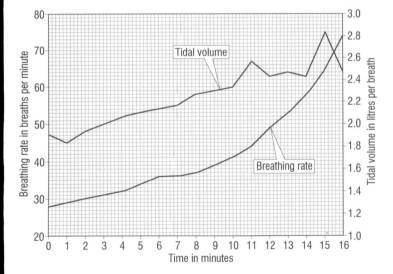

The graph shows the effect of the exercise on the athlete's breathing rate and tidal volume.

  a **i** What is meant by tidal volume?   (1)

    ii What is the athlete's tidal volume at 3 minutes?   (1)

    iii The difference in breathing rate from 0 to 8 minutes is 9 breaths per minute. Calculate the difference in breathing rate from 8 to 16 minutes.   (1)

    iv Explain why breathing rate increases as the speed of the running machine increases.   (4)

  b The diagram shows how the lungs are ventilated.

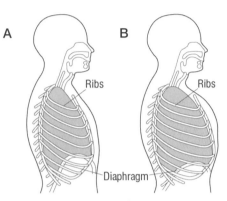

Which diagram, A or B, shows inhalation (breathing in)? Explain your answer.   (3)

AQA, 2010

5 An athlete needs glucose and oxygen for energy in the muscles. More oxygen is needed when exercising.

  a **i** Describe the changes in the body that would increase the amount of oxygen delivered to the muscles.   (3)

    ii Name the process that provides energy to the muscles if there is not enough oxygen.   (1)

    iii Explain why this process releases less energy than if there was a plentiful supply of oxygen to the muscles. **[H]**   (3)

  b Athletes are advised to eat more complex carbohydrates, such as pasta, the night before a race. Explain why.   (3)

AQA, 2010

1   Sports nutritionists give athletes advice about how to improve their fitness. BMI values give an indicator as to whether someone is the correct weight for their height. An athlete had their measurements taken to get advice about their fitness.

   Mass = 108 kg   Height = 196 cm

   a   What does BMI stand for?   *(1)*

   b i   Use the results given to calculate the BMI for this athlete.   *(2)*

   ii   Sports nutritionists can use the table below to show whether or not an athlete is a healthy weight. Use the table to give a rating for this athlete.

BMI	Rating
Less than 18.5	Underweight
18.5–24.9	Normal
25–29.9	Overweight
30+	Obese

                                                                                          (1)

   iii   Suggest why the BMI rating for this athlete could be misleading.   *(2)*

   c   Another athlete has a BMI of 22.5. She asks the nutritionist for advice on how to build muscle. Suggest what nutrient this athlete should increase her intake of.   *(1)*

2   An athlete is training for a weight lifting competition. She needs to develop her arm muscles. The athlete's sports physiologist gives her some exercises to do on a chinning bar.

   a   Look at the diagram of the muscles in the arm.

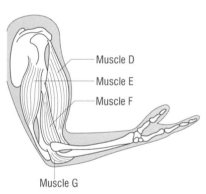

   Muscle D
   Muscle E
   Muscle F
   Muscle G

   i   Which muscle, **D**, **E**, **F** or **G**, would contract to lift the athlete's body?   *(1)*

   ii   Which muscle, **D**, **E**, **F** or **G**, would contract to lower the athlete's body?   *(1)*

   iii   Name muscle **D**.   *(1)*

   iv   What are *antagonistic* muscles?   *(1)*

   b   After a short time on a chinning bar, the athlete noticed some changes in her body. Choose **three** changes from the box below that the athlete would notice in her body after a short time on the chinning bar?

Heart rate increases   **Urination increases**   **Skin becomes redder** **Skin temperature rises**   **Breathing rate decreases**

                                                                                          (3)

AQA Examiner's tip

This is a calculation question. It is important to show the equation you have used and all your working out even if you have worked it out on a calculator. This means you should show how you have substituted the numbers into the equation. There are 3 marks for this question so you will get all 3 marks if you get the correct answer. However, if you make a mistake you can still get a mark for the equation and a mark for the working out.

AQA Examiner's tip

If a question asks you to use the information in the table it is important that you do so even if you think there may be another way to answer the question.

c The equipment shown in the diagram is used to measure the strength of the athlete's muscles.

Describe how to use this equipment to measure the athlete's muscle strength. *(3)*

d During training the weight lifter damages her elbow joint.

  i Name the fluid surrounding the cartilage in the elbow joint. *(1)*

  ii Describe and explain the reason for the weight lifter feeling pain in her elbow. **[H]** *(2)*

AQA, 2008

3 A sports physiologist has to assess the basic health and fitness of an athlete. He listens to the athlete's heart through a stethoscope. He must be familiar with the structure and the working of the heart.

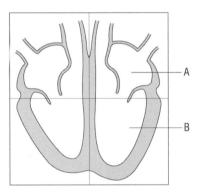

a Name the chambers of the heart labelled A and B *(2)*

b When the athlete is exercising more oxygen is needed in the muscles than when they are at rest.

  i Why is more oxygen needed in the muscles during exercise? *(1)*

  ii Which part of the blood carries the oxygen to the muscles? *(1)*

c i Aerobic respiration is a process occurring in the muscles that provides the energy needed for the athlete to carry out exercise.

    Copy and complete the word equation for aerobic exercise:

    Glucose + oxygen $\longrightarrow$ *(1)*

  ii When an athlete has been exercising for a while, their body may need to provide energy through the process of anaerobic respiration rather than by aerobic respiration. Suggest why anaerobic respiration sometimes occurs. *(2)*

4 *In this question you will be assessed on using good English, organising information clearly and using specialist terms where appropriate.*

A sports nutritionist has to consider how diet can help maximise an athlete's performance. Some athletes use a high protein diet. Some use a high carbohydrate diet. Explain the purpose of each of these types of diet and compare them with each other. You should consider the advantages and the disadvantages of each diet for an athlete. *(6)*

The use of science to develop materials for specific purposes

Making connections

In this chapter you will learn about how and why materials scientists choose materials for particular purposes. You will look at the properties of materials, how they are tested and what those materials are used for.

Figure 2 Metal corrosion research

Figure 3 Spacecraft material design

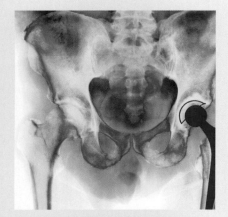
Figure 7 Titanium alloy for hip joint replacement

Figure 1 Materials scientists are changing our world

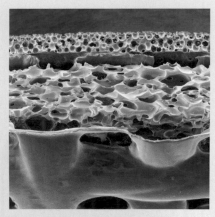
Figure 6 Polyurethane wound dressing

Figure 5 Sports footwear investigation

Figure 4 Tensile strength testing

42

Advances in materials technology

Materials scientists help design and make new products using many different materials. They are constantly researching, developing and testing new materials. Economic, energy, environmental, health and manufacturing issues influence their work.

The use of iron gave birth to Britain's Industrial Revolution. Although iron is still important in the construction and car industries, the use of new materials is shaping our future. The cost to industry of preventing iron from rusting is huge – no wonder alternatives such as aluminium and plastics dominate today's consumer market.

Practical

Resistance to corrosion

When iron rusts it forms iron oxide. This flakes off the iron or steel. The industrial cost of corrosion is huge.

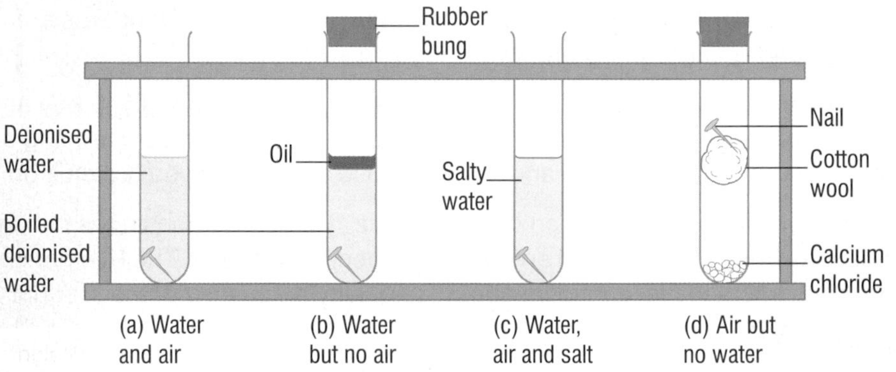

Figure 8 The corrosion resistance test

Prepare four sets of test tubes as shown to compare iron, stainless steel (an iron alloy containing 11% chromium), galvanised (zinc-coated) nails and a piece of aluminium metal. (A total of 16 test tubes will be needed.) The oil in tube b stops air reaching the nail. The calcium chloride in tube d removes water from the air.

Leave for a week before noting changes in appearance.

● Which tubes show no sign of rusting?

● What chemicals are essential for iron to rust?

● How can rusting be prevented?

Safety: Anhydrous calcium chloride is an irritant. Wear eye protection.

3.1

Introduction to materials science

?? ? Did you know ... ?

There are 27 000 current British standards. That's one for every 59 businesses in the UK.

Figure 1 BSI Kitemark symbol

Figure 2 CE marking

⊙⊙ **links**

For more information on the standard procedures to test materials for density see 3.8 Standard procedures for testing materials.

Fitness for purpose

Before going on sale products and materials are quality-tested. This is to ensure they are safe to use and meet national and international standards. Companies, by law, must know of no reason why their product is not satisfactory for the job (fit for the purpose) for which it is designed.

Standards

A standard is a guide – a rule if you like – that all companies who design and make things should follow for good practice. These guidelines or standards are set by The British Standards Institution (**BSI**), the European Committee for Standardisation (**CEN**), or the International Organization for Standardization (**ISO**). They make sure that products are safe to use and more reliable. European Standards are adopted as British Standards. British Standards are codes of good practice or agreed ways of doing something.

Many things you buy conform to British Standards. For example, if you buy an electrical appliance, the 'standard' plug fits into a 'standard' socket, sending a current down a 'standard' wire, and it won't cause a fire when you switch it on.

Many products display a 'Kitemark®' (see Figure 1) which demonstrates that the product has been tested by BSI and meets a specific standard. The Kitemark® allows consumers to feel confident about the quality and safety of the product.

The '**CE** mark' (see Figure 2) is needed by many products in order for them to be bought and sold in the European Union. CE stands for *conformité européenne*, French for 'European conformity'. By placing the CE mark on a product, a manufacturer is stating that is meets the requirements of the appropriate European Directive, ensuring that the product is well designed and safe for the user.

> **a** What is meant by the phrase 'fit for purpose'?
>
> **b** What is a British Standard?

Measuring material properties

When scientists look to develop products like sports gear or protective equipment, they must consider the materials it will be made from. They will think about:

- What properties does the product need for the job it does?
- Which materials have the most suitable properties for this application?
- How should the article be designed with the chosen material?

Scientists describe materials by their properties, which they measure in a standard way all over the world. Properties include **density**, electrical conductivity, strength, hardness and flexibility. The way we use a material depends on its properties, and these properties are affected by its internal structure.

Density

What a material is used for often depends on its density. For example, many items of sports equipment, such as tennis rackets, use low density material even when it is relatively expensive. There are British Standards on determining the density of materials.

Steel has three times the density of aluminium. The same volume of steel has about three times the mass of aluminium. Density is measured in units of g/cm³ using the formula:

Density (g/cm³) = $\dfrac{\textbf{mass (in g)}}{\textbf{volume (in cm}^3)}$

If an object has a density greater than that of water (>1 g/cm³) it sinks.

A force accelerates an object with less mass more quickly. That is why Formula 1 racing cars use low density materials.

> **c** Use volume = length × width × height to calculate the volume of the steel and aluminium blocks in Figure 3.
>
> **d** Use the formula: density = mass/volume to show the density of **i** steel is 8.0 g/cm³ and **ii** aluminium is 2.7 g/cm³.

Electrical conductivity

We use good electrical conductors like copper to carry currents in homes, cars and factories. A material that resists an electric current passing through it has a poor conductivity. A longer wire has proportionally more resistance than a shorter one.

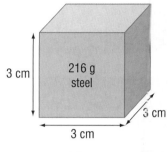

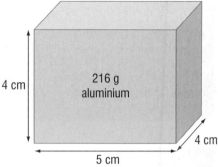

Figure 3 Steel and aluminium blocks with the same mass

Practical

Electrical conductivity

Measure the resistance of 1 m of thin copper wire using a multimeter set on the Ω scale.

Compare to another copper wire of the same diameter that is 2 m long.

Compare to another material 1 m long of the same diameter.

Their conductance = 1 ÷ resistance (in Ω⁻¹)

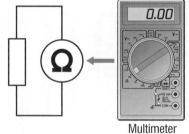

Multimeter

Figure 4 Using a multimeter

??? Did you know ...?

Sprint shoes need to be light and rigid. Spikes on the shoes increase grip. By using low density materials the shoes are lightweight. The less the mass of the shoe the easier it is to increase speed.

Copper is almost twice as good at conducting electricity as aluminium and ten times better at conducting than steel. Copper is more than three times as dense as aluminium. Neither copper nor aluminium by itself is suitable for overhead power lines. Both stretch easily and would start to sag if hung on pylons or over rail tracks. Steel, by contrast, is strong and stiff. If we strengthened copper with steel, the wire would be heavy and we would need huge structures to support it. However, aluminium wire with a thin but strong steel centre is both light and good at conducting, so is favoured for overhead power lines.

Summary questions

1 What does a 'CE marking' mean to the consumer?

2 What is the density of a copper rod of mass 45 g and volume 5 cm³?

3 The resistance of 1 m of copper wire with cross-sectional area 1 mm² is 0.02 Ω. What is the resistance of a wire with the same diameter that is 100 m long?

Key points

- 'Fitness for purpose' implies 'satisfactory for the task intended'.

- The British Standards Institute (BSI) and European Committee for Standardisation (CEN) set and test product standards.

- Density = mass ÷ volume (units: g/cm³)

- Good conductors do not resist the flow of electrical current.

3.2 Forces on materials

Compression and tension

Forces are measured in newtons (N). A force is either a push or a pull. A tensile force or **tension** is a pull that stretches an object, while a compressive force or **compression** is a push that squashes. When objects bend, one part is in tension while another part is in compression, for example in sailing masts and golf shafts. The outer edge stretches, the inside compresses.

a What is the difference between a tension and a compression?

Hooke's law

When metals, made into springs or wires, stretch they follow a pattern, known as **Hooke's law**. Add weights to a spring and you can see that 'the extension is directly proportional to the force applied'. This means that if you double the force then you will double the extension. When the weight is removed, it springs back to its original length. Add too many weights, however, and it will no longer spring back. The spring has now passed its **elastic limit**.

Figure 1 Bending a mast causes tension and compression

Maths skills

Springs and Hooke's law

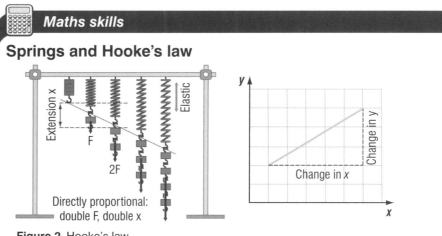

Figure 2 Hooke's law

The gradient of the force–extension graph shows the stiffness of the spring.

The gradient (or slope) of a straight line shows how steep the line is.

Gradient = change in y ÷ change in x

This gradient is called the **spring constant** k, measured in N/cm. The green graph line shows a spring with constant:

$$k = 1\,N ÷ 2\,cm = 0.5\,N/cm$$

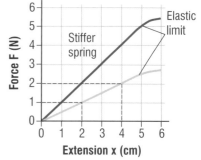

Figure 3 The gradient of a slope

Hooke's law: Force = constant × extension, or $F = k \times x$

  (N)   (N/cm)   (cm)

so when this equation is rearranged the spring constant, $k = F ÷ x$ (in N/cm).

b i What does the spring constant describe?

  ii What is the constant of the spring shown by the blue graph line?

c If 12 N stretches a spring by 8 cm, how much will it extend if 3 N is applied?

Tensile breaking strength and stress

Wires, as well as springs, also obey Hooke's law, up until their elastic limit. Applying more force will eventually break the wire. A wire with a high **tensile strength** resists stretching. (Materials with high **compressive strength** resist crushing or squashing.) Steel cable, to support bridge structures and ships' anchors, needs to be both corrosion-resistant and strong, with a high tensile strength.

d Wires in tension also support dental braces. Give two other uses for wires or cables in tension in transport and medicine.

Practical

Changing the properties of steel

Steel is an alloy (or mixture) of iron, with other elements added to it. Adding up to 1.5% carbon changes how steel fractures, and increases the strength of steel wire. Adding 3% carbon makes it brittle and weaker.

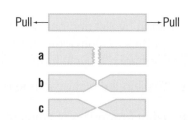

You can model this by comparing how Plasticine breaks when fine sand is mixed with it.

Figure 4 Plasticine fracture with fine sand added **a** 20%, **b** 10%, **c** no sand

Brittle materials crack, leaving sharp edges or fragments. **Tough** materials are not brittle. They absorb energy and deform by necking before they break.

● What variables will you need to control?

Scientists often measure the tensile breaking strength of a wire not in units of force (N), but in units of **stress** (N/mm²). The formula to measure stress is:

$$\text{Stress (N/mm}^2) = \frac{\text{force (newtons, N)}}{\text{cross-sectional area (mm}^2)}$$

Figure 5 Cross-sectional area = πr^2

The cross-sectional area of a wire = $\pi \times r^2$ (where r is the radius of the wire in millimetres).

e The diameter of a stainless steel cable is 10 mm.

i What is its radius?

ii What is its cross-sectional area?

f The cable breaks when a force of 62 800 N is applied. Show that its tensile breaking strength is 800 N/mm².

links

For more information on the standard procedures to test materials for toughness/brittleness see 3.4 Composites, wood and ceramics. For tensile breaking strength see 3.8 Standard procedures for testing materials.

Summary questions

1 Draw the forces on a golf club shaft that bends as it hits a ball.

2 **a** Why do Airbus use a strong, low density aluminium–titanium mixed alloy rather than steel for their aircraft?
b Why does a strong, low density, titanium alloy bicycle frame benefit riders?

3 A spring has a constant of 0.25 N/cm. What force extends the spring by 4 cm? (Assume the elastic limit is not exceeded.)

4 The tensile **breaking stress** of a particular **steel** is 1200 N/mm². What force will be needed to **break** a wire with a diameter of 0.5 mm?

Key points

● Tensions are pulls that stretch; compressions are pushes that squash.

● Hooke's law: Extension is directly proportional to applied force.

● Stress = force ÷ area (units: N/mm²)

● Brittle materials crack easily. Tough materials absorb energy and stretch before breaking.

3.3 Metals and alloys

Learning objectives

- What is an alloy?
- What are the properties of metals and their alloys?
- Why are metal alloys used in sports, medical and transport applications?

∞ links

For information on composites, wood and ceramics see 3.4 Composites, wood and ceramics and for polymers see 3.6 Polymers. Refer to 5.2 Distinguishing different chemicals for more details about metal ions and see 3.8 Standard procedures for testing materials, for details about how to measure hardness and stiffness/flexibility.

Types of material

There are five important types of material used in manufacturing: metals, composites, wood, ceramics and polymers. Their different properties make them suitable for different applications.

An **alloy** is a special composite made from two or more elements, at least one being a metal.

Aluminium–titanium alloy is as strong as steel, but much less dense, making it far more suitable for aircraft frames. Light alloy aircraft need less fuel to lift them off the ground and fly.

Although the first replacement hip joints were made from steel, strong, low density, titanium alloy has the advantage of having a density similar to that of the bone it is replacing.

> a A racing cyclist can easily accelerate a lightweight bicycle. What properties of aluminium–titanium alloy make it suitable for a high performance bicycle frame?
>
> b i Explain the disadvantage of using steel for a replacement hip joint.
>
>   ii What properties of titanium alloy make it suitable for a hip joint?

Properties of metals

Metals such as steel, titanium and aluminium alloys all have properties that are useful. Metals are:

- **strong** with a high tensile strength (needing a large force to break them)
- good **electrical** and **thermal conductors**
- **stiff** (not **flexible**)
- **hard** (difficult to dent or scratch)
- **malleable** (can be hammered into shape and rolled into sheets).

> c Suggest some examples of where metals are used for transportation, medical applications and sports equipment and why these metals are suitable.

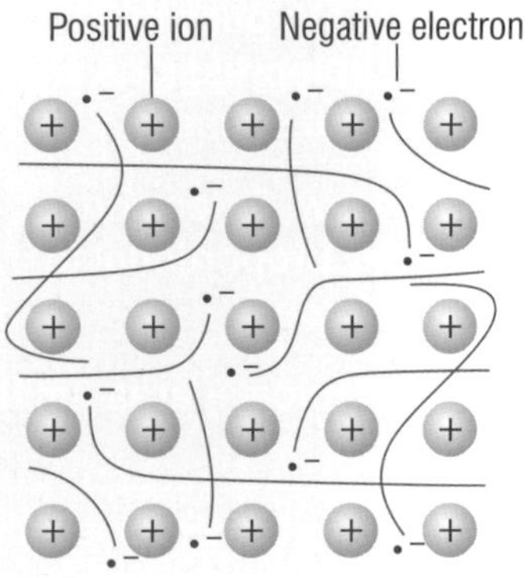

Models for metals

All atoms consist of a positive **nucleus**, surrounded by negative electrons. Unlike other elements, the outer electrons of metals are not attracted very strongly by their nucleus, in fact these electrons can flow from atom to atom. This explains why metals conduct electricity so well. The movement of outer electrons also helps to make metals good conductors of energy.

A metal atom that has lost an outer electron is called a positive **ion**. Think of the structure of a metal as an arrangement of positive ions 'in a sea' of negative electrons. The ions and electrons attract each other strongly. It is hard to break the bonds that hold them together. This explains the strength and hardness of metals.

Positive ion   Negative electron

Figure 1 Metallic bonding – an arrangement of positive ions in a sea of negative electrons.

Practical

Bubble raft model

Safety: Keep room ventilated and avoid sharp syringes.

With the syringe in the bubble mixture, limit the gas flow with the Hoffman clip.

Move the syringe back and forth to produce a sheet of bubbles and switch off the gas.

Notice the surface pattern and dislocations (see Figure 3).

Place the ends of two rulers into the bubbles, about 5 cm apart.

Gently move them together and apart to squash and stretch the pattern of bubbles.

Notice that the pattern first moves where there are dislocations.

Notice that the presence of some larger bubbles tends to hold the structure together.

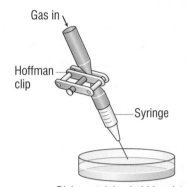

Figure 2 Making bubbles

Steel

Steel is an alloy of iron containing carbon and various other alloying elements, like manganese, chromium, vanadium and tungsten. Adding more than 11% chromium stops steel rusting. The other elements harden and strengthen the iron by preventing dislocations in the structure from sliding past one another.

Steel's strength has allowed the building of skyscrapers and oil platforms. Before steel, the only way to build high was to pile up rocks! Its strength makes it ideal for structures like car frames and ploughs. Its hardness makes it perfect for sharp knives and surgeon's scalpels. Apart from concrete and stone, industry turns out more tons of steel than any other material.

d What properties of steel make it suitable for surgical instruments?

Summary questions

1 What do these words mean? **a** hard **b** stiff **c** strong

2 Explain why metallic bonding makes metals both strong and good electrical conductors.

3 Explain why W. L. Bragg thought that creating an alloy might make a metal stronger.

Scientists @ work

Bubble raft

In 1947 Nobel Prize winner W. L. Bragg suggested using a 'bubble raft' model to represent the metallic bond and explain the behaviour of metals and alloys under tension or compression. He hypothesised that, like the bubbles in the dish, metals may not have a regular structure throughout. There may be imperfections where bubbles don't line up properly. Some metals may not be as strong as others if they have imperfections (or gaps) in their structure. Creating an alloy, by adding different sized atoms into the structure, may fill the gaps, stop the structure slipping, and make the metal stronger.

Dislocations or gap   Alloy filling gap

Figure 3 Dislocations and alloying

Key points

- An alloy is a mixture of two or more elements, at least one being a metal.

- Metals and their alloys are strong, malleable and good electrical and thermal conductors.

- Strong titanium alloy for racing bicycles has a low density.

- Stainless steel surgical instruments are strong and hard. Low density, yet stiff, strong aluminium alloys are ideal for aircraft frames.

3.4 Composites, wood and ceramics

Learning objectives

- What are composites and ceramics?
- How do composites benefit from the properties of their component materials?
- What are the properties of composites, wood and ceramics that make them suitable materials to use?

Composites

Composites contain a mixture of different materials that bond together. By combining materials intelligently scientists can produce a mixture with better properties than its component parts.

A tough material is not brittle. Glass is strong but brittle. Plastic is weak but tough. Adding glass fibres to plastic makes a strong, tough composite ideal for boat hulls.

Carbon-fibre–plastic composites have a low density, but high strength. Besides their use for bicycle frames and tennis rackets, sports materials scientists use carbon-fibre to manufacture badminton rackets, racing dinghies and yachts, and the shafts of golf clubs. Although carbon-fibre composites are strong they can break due to a severe impact. Titanium metal is strong and has a low density. The cockpit of a motor racing car is reinforced by a composite containing a mesh of titanium and carbon fibres.

> **a** Use an example to explain how a composite material can have better properties than its component materials.
>
> **b** Why are titanium–graphite fibres used in tennis rackets?

Wood

Wood is a natural composite containing cellulose fibres embedded in lignin. Cellulose has tensile strength and lignin resists compression, making wood a strong, lightweight material. Although used mainly in the construction of houses and furniture, many types of sports equipment are produced from wood. The ideal wood for cricket bats is White Willow. Besides being strong and light, it is tough and shock-resistant, so the wood dampens vibrations when you hit the ball. In 2010 scientists announced that wood could even be used as a future bone substitute.

> **c** Ash is a dense, hard, tough, strong, elastic wood. Explain which two of these properties make it most suitable for archery bows.

Ceramics

Ceramics are inorganic, non-metallic materials, e.g. pottery (made of clay) and glass. Ceramics:

- are hard
- are brittle (can crack easily and are not tough)
- have a high melting point
- have a low thermal conductivity
- resist chemical attack.

> **d** The space shuttle collides with air molecules at a speed of 7 km/s as it re-enters the Earth's atmosphere. What properties of heat-resistant ceramics make them suitable for the outer tiles on the space shuttle?

Scientists @ work

Tennis rackets

The key properties of a tennis racket are stiffness, light weight and vibration dampening. After 700 years the wooden tennis racket became redundant. Wood's flexibility, or lack of stiffness, limited power in play.

In 1974 Jimmy Connors won the Wimbledon and US Open finals using an all-steel tennis racket. Steel did not dampen vibrations, so added to repetitive stress injuries in muscles and tendons.

In 1980 John McEnroe began four years as world no.1 using the graphite Dunlop Max 200 g. It was made of carbon fibres bound together with a plastic resin. 1998 saw the introduction of the Head Ti6S. This was a mixed composite racket, with strong titanium and light graphite fibres woven together. Today some manufacturers add Kevlar to increase the racket's ability to absorb the shock of vibrations.

Car engines emit toxic chemicals from their exhausts. Catalytic converters change some of these into non-polluting gases.

e Explain why the ceramic's heat and chemical resistance properties are important in a catalytic converter.

Testing for toughness

Scientists can use the Izod test to compare the toughness of different materials and how the material fractures on impact.

Figure 1 Carbon–ceramic composite disc brakes

Practical

Toughness/brittleness – the Izod impact test

Safety: Avoid being hit by the hammer or broken fragments. Wear eye protection.

Select three similar sized samples of aluminium alloy, plywood and ceramic tile.

Arrange the apparatus as shown, with a notched sample of material clamped in the vice.

Release the hammer at increasingly steep angles until the sample breaks.

Note in a table the angle of release and the type of fracture (extent of cracking or necking) to determine the strength of each material and its brittleness or toughness.

Repeat with another material.

● Describe the differences between tough and brittle materials when they break.

● Why is it important in the Izod test to:

 a have materials of the same size and thickness,

 b make the V-shaped notch in each material the same depth?

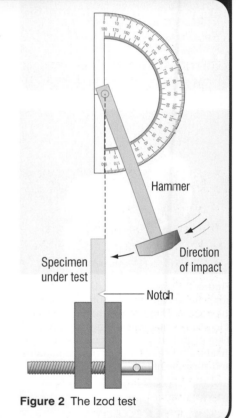

Figure 2 The Izod test

Summary questions

1 Classify each of the following as alloys, composites, wood or ceramics:
 a a brick
 b a willow cricket bat
 c an aluminium–titanium mast
 d a glass-fibre-reinforced plastic boat hull.

2 Why is it safer to take advantage of tough rather than brittle materials for sporting and transport uses?

3 State two composite and ceramic materials used in sporting and transport applications and explain why these materials are fit for purpose.

4 **a** What properties are important in the design of a tennis racket?
 b How does adding Kevlar help make carbon composite tennis rackets and the shafts of golf clubs better?

Key points

● Composites contain a mixture of two materials. Composites benefit from the combined properties of their components.

● Wood is strong, with a low density.

● Ceramics are non-metallic materials like pottery and glass. Ceramics are hard and brittle, have a high melting point and resist chemical attack.

3.5 Natural or synthetic?

Learning objectives

- What are the benefits and drawbacks of natural and synthetic materials?
- Why are natural or synthetic fabrics chosen in sports, medical and transport applications?
- What affects the thermal conductivity of fabrics?

Figure 1 Silkworm cocoons. Silk is a soft, strong natural fibre obtained from the cocoon of the silkworm. After soaking it in soapy water you can unwind up to 1 km of fibre from the cocoon. Its strength makes silk useful for clothing, parachutes and surgical thread.

Figure 2 Policewoman wearing Kevlar body armour

Look at your shirt label. What is the shirt made from? We make clothes using either **natural** or **synthetic** (man-made) polymers or a mixture of both. Natural fibres, like cotton, silk, wool and leather, come from living things. Synthetic fibres, like polyester and polyurethane Lycra, are man-made and are often produced from chemicals we get from crude oil. Lycra can stretch up to seven times its original length without breaking, but springs back elastically to its original shape. It is usually mixed with fibres like cotton or polyester. Tight-fitting, yet flexible, Lycra body suits and medical support socks encourage blood flow round the body to the muscles. Polyurethane fibre is also used as the mesh on sterile plasters. In combination with the absorbent cotton pad, the plaster keeps the wound clean and dry.

a Explain an application in **i** sport, **ii** medicine where polyurethane is appropriate?

The table below lists the advantages and disadvantages of natural and synthetic fibres.

Fibre	Advantages	Disadvantages
Natural	• obtained without much chemical processing • biodegradable (decays naturally) • absorb water, so are comfortable	• limited number available • ignite easily • can shrink in the wash
Synthetic	• inexpensive to make • strong and hard-wearing (resistant to abrasion, creasing and staining) • can be brightly coloured • dry quickly after washing	• low melting point, melt to skin if burnt and produce toxic fumes • non-biodegradable • can feel clammy and stick to skin, so limiting breathability

A polyester cotton blend is an almost ideal clothing fabric. The cotton absorbs moisture, which makes for comfort, while the polyester is more hard-wearing and crease resistant than if the garment were made from cotton alone.

b Which type of fibre is more resistant to abrasion (wear and tear)?

c **i** Suggest why a garment made from a synthetic polymer, like nylon, is uncomfortable to wear in the summer.

    ii Why would a natural fibre one be better?

d Why is polyester perfect for people who hate ironing?

Kevlar

Kevlar is a lightweight, tough synthetic fibre used for the lining of bicycle tyres and to replace damaged ligaments. By weight, Kevlar is five times stronger than steel.

Bullet-proof police vests benefit from layers of Kevlar fibres woven together. The fibres spread the force of a bullet over a large surface area, distort the bullet and absorb its kinetic energy. In addition, a layer of resin hardened glass fibre between the Kevlar layers prevents knife blades entering the jacket. For extra protection, military jackets sometimes contain sheets of titanium between the Kevlar layers.

Mixed with carbon fibres and epoxy resin, a Kevlar composite is ideal for strong yet energy-absorbing tennis rackets and golf club shafts.

e State two uses for Kevlar and explain why the material's properties uniquely suit each of these applications.

Shape affects properties

Articles of clothing with a larger surface area have more surface molecules to absorb moisture. They also dry quicker. You do not scrunch up clothes to dry them! Thickness also affects a material's absorbency. Moisture-absorbent padding helps to take water away from sweaty feet in trainers.

Objects with a larger surface area cool down quicker. Thicker materials are better for thermal insulation and maintaining body temperature on a cold day. Thermally insulating clothing works by trapping air between fibres. The more air that stays trapped in your sleeping bag the more difficulty the molecules will have in conducting energy away from your body.

Practical

Comparing the thermal conductivities of different fabrics

Sandwich your fabric sample between an acetate sheet and a thermochromic sheet.

As soon as your beaker of water boils, stop heating and place the sandwich on top.

Measure the time it takes for the thermochromic sheet to just change colour.

Optional: Take surface temperature readings every 10 s with an infrared thermometer.

- Which fabric is the best insulator?
- Which is the best conductor?

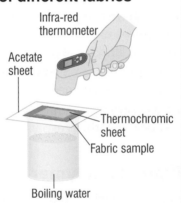

Figure 3 Comparing the thermal properties of fabrics

Safety: Take care with hot surfaces and boiling water. Wear eye protection.

Summary questions

1 a Name two natural fibres and two synthetic fibres.
  b Give a use for each of these examples.
  c State the properties of each material that make them fit for the uses that you have given.

2 Lycra clothes are both flexible and tight-fitting. How can Lycra help in
  a athletics b medicine?

3 Explain why Kevlar is so useful:
  a for replacement ligaments
  b in bullet-proof jackets
  c as a composite material in golf club shafts.

4 Explain how a mountaineer could benefit from wearing clothes containing 'phase change material'.

Scientists @ work

Phase change material

A special material called 'phase change material' can absorb and release energy, keeping the body temperature constant. If the body temperature of an athlete rises during exercise, the gel in the material absorbs energy and melts. If the athlete's temperature falls, the gel refreezes and releases the energy that was absorbed.

This makes phase change material useful for ski gear and clothing for mountaineers, whose body temperature would otherwise rise with exercise, but fall during rest on a cold mountain-side.

It can also maintain a steady cool temperature inside potentially hot gloves, helmets and trainers.

∞ links

See 3.8 Standard procedures for testing materials.

Key points

- Natural fibres absorb water, but ignite easily. Synthetic fibres are hard-wearing, but can feel clammy.

- Consider the advantages of the properties of a material for a particular application, such as light weight (faster accelerations), durable (long-lasting), flexible (for comfort), waterproof, insulating, breathable (letting air through) and moisture absorbing.

3.6 Polymers

Learning objectives

- What are polymers, and what are their typical properties?

- How does chain length and branching affect the properties of a polymer? [H]

- What are thermoplastic and thermosetting polymers? [H]

Figure 1 Polymers have many uses. Hard plastics and hot soft glue are used in this glue gun.

Where are polymers used?

'Poly-mer' means 'many-bits'. Polymers are organic compounds with long-chain molecules. The chains are made by thousands of small molecules (called monomers) joining together in a chemical reaction. There are strong covalent bonds between the atoms in each chain and weaker forces of attraction between the separate chains. Wool and cotton are natural polymers. Plastics like polyvinyl chloride (PVC – used for cling-film), polyester (used for clothing) and nylon (used for rope and fishing line) are synthetic polymers.

Polymers:

- are flexible (so are not stiff)
- have a low density
- have a low thermal conductivity.

> **a** Lorry drivers tie pallets down with nylon rope. What properties of nylon make it fit for this purpose?

Although we see plastics polluting our environment, biodegradable alternatives are more costly and denser. Instead of dumping old plastic drinks bottles in landfill sites, we can recycle them.

- Tough, flexible polypropene (PP) is used for cups and bottles. Medical uses include replacement valves and joints, and even hernial patches.
- Hard polystyrene (PS) is used for car number plates and Petri dishes, while shock absorbing expanded polystyrene (EPS) is used for packaging.

Types of polymer

You can model the behaviour of different polymer chains using spaghetti. The spaghetti strands behave like long-chain molecules. The different way that stickier or shorter spaghetti strands respond can be used to model differences in properties between different polymers.

The properties of polymers are governed by their chain length, the amount of branching and the strength of the forces between their chains. These affect the polymer's strength, density and melting point. Scientists can add chemicals to polymers to create intermolecular bonds, called **cross-links**, between the chains. The degree of cross-linking determines the final properties of the material.

Thermoplastic (thermosoftening) polymers

The most widely used plastic is polyethene. It is a **thermoplastic polymer** and has few, if any, cross-links between its long chains. Its relatively weak intermolecular bonds allow it to deform and flow when put under tension. A thermoplastic polymer is flexible and softens when heated, so is easy to mould and shape.

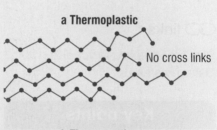

a Thermoplastic

No cross links

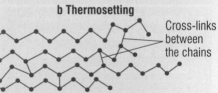

b Thermosetting

Cross-links between the chains

Figure 2 Altering the chain length, the amount of branching and cross-linking affects the properties of a polymer

Higher

Practical

The spaghetti hypothesis

Run a fork through two bowls of spaghetti; one where the strands are loose and separate, the other where they are sticking together.

Describe the effect of:

● the strands (polymer chains) being intertwined or 'entangled' with each other

● the strands sticking together (the chains being more strongly attracted to each other).

Cut the spaghetti strands up into shorter lengths in each bowl.

● What effect does having shorter strands (chains) make?

If the hypothesis is valid, in what type of polymer should it be:

● more flexible? Is it one with weaker or stronger forces between the chains?

● more dense? Is it one with weaker or stronger forces between the chains?

● easier to separate the chains from one another, so giving it a lower melting point?

Thermosetting polymers

A **thermosetting polymer**, like vulcanised rubber, has strong cross-links between its polymer chains. Additives like sulfur modify the properties of natural rubber by forming cross-links between the individual polymer chains. The newly formed vulcanised rubber is less sticky, denser and harder than the original natural rubber. Its uses range from car tyres and shoe soles to ten-pin bowling balls.

Cross-links make thermosetting polymers rigid once set, and cause them to decompose, rather than melt, when heated. Thermosetting polymers cannot be remoulded.

b Plastic mouth guards for ice-hockey and rugby players soften in hot water so they can be moulded to fit the player's mouth. Which type of polymer is likely to be used? Explain.

c How do scientists create very strong dense polymers like ten-pin bowling balls?

Demonstration

The plastic cup trick

With caution, heat a plastic 'vending machine' cup with a hot air gun or hair drier and watch it shrivel to its original pre-stretched state.

Safety: Do not touch the sides of the hot hair dryer or the hot cup.

Key points

● Polymers are organic compounds with long-chain molecules. Polymers are flexible and have low densities and low thermal conductivities.

● Irregular branching means chains cannot pack together as well as straight-chain polymers. This results in lower density polymers which melt at lower temperatures. **[H]**

● Increasing chain length causes more entanglement.

● Cross-links strengthen the forces between chains. These increase a polymer's density and make it harder to melt. **[H]**

● Polymers are long-chain molecules. Weak intermolecular bonds between chains allow moulding of thermoplastic polymers. Strong cross-links in thermosetting polymers do not allow remoulding. **[H]**

Summary questions

1 What does 'polymer' mean?

2 Polyethene is a thermoplastic polymer.
  a What happens when it is heated? **[H]**
  b What advantage does this create, compared to a thermosetting polymer? **[H]**

3 **a** How do cross-links create thermosetting polymers? **[H]**
  b What advantages does this create, compared to a thermoplastic polymer? **[H]**

3.7

Materials for sports, medicine and transport

Advances in materials technology

Only by understanding the properties of materials can scientists improve the effectiveness of manufactured products. With the development of so many new materials, no wonder every year sports records get broken, advances are made in medicine and improvements take place in vehicle design.

Bicycle frames

The design of the bicycle, with its solid frame and equal-sized wheels, is basically the same as it was 100 years ago. Before the 1980s bike frames were made from steel, which is easily manufactured, inexpensive and strong, but heavy. However, in the 1980s, the use of expensive new materials, such as low density titanium and carbon fibre, allowed designs to change.

A revolutionary aerodynamic design helped Chris Boardman win a 1992 Olympic gold medal. His Lotus bicycle had a strong composite shell. The lightweight frame, made of carbon fibre in an epoxy resin, was strengthened with titanium at places of high stress.

a What is a disadvantage of: **i** steel? **ii** carbon-fibre composite?

Figure 1 Chris Boardman's LotusSport Pursuit bicycle

Bicycle helmets

From cycling's earliest days there were head injuries. Until the 1970s racing cyclists wore strips of leather padded with wool or sponge, but these had only limited shock-absorbency.

The first mass produced helmet with shock-absorbing expanded polystyrene or **EPS** went on sale in 1974. In a crash the air pockets within the EPS compress and absorb kinetic energy, avoiding the dangers of a sudden impact. However, to hold the EPS together, this helmet had a heavy hard plastic shell. In 1986 the heavy supporting shell was replaced by a Lycra cloth and the foam was reinforced with a strong nylon mesh.

The next big design step, in 1990, was the re-introduction of a thin smooth cover. This time in low density milk carton plastic or **PET** (polyester). Being smooth and round, these helmets skid more easily, and reduce the likelihood of jarring head injuries. Moulding the foam inside a shell also helps to hold the foam together in an impact. While the outer shell spreads the impact, forming many tiny cracks, the inner EPS foam is crushed and absorbs the force.

Since Greg LeMond won the *Tour de France* with an aerodynamic design, elongated helmet shapes with cooling air vents, rather than rounded ones, have become the fashion.

Activity

Research the evolution of the bicycle frame.

 Did you know ...?

Invented in 1981, laminated glass windscreens have a thin vinyl plastic layer sandwiched between two layers of glass. If the windscreen breaks, the glass does not shatter.

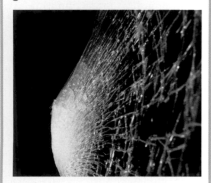

Figure 2 Car windscreen testing

b What happens to the EPS in a cycle helmet when it absorbs forces in a crash?

c Explain how allowing a helmet to skid more easily and moulding the foam improve safety.

d EPS is a good insulator. What modern bicycle helmet design feature prevents a cyclist's head from getting too hot?

Medical materials

The properties of materials critically affect their performance – properties such as density, strength, smoothness, flexibility, shock-absorbency and insulation. For example, implant materials must be chemically biocompatible with our bodies so we do not reject them, and be able to handle stresses so they do not break. Nowadays biomaterials scientists design medical devices using synthetic polymers and composites (combinations of different materials) to make some very smart materials. The density of tungsten is 70% more than lead, so a tungsten-filled, flexible polymer composite is ideal for clothing to shield radiographers from X-rays.

By combining polymers with mammal cells it is now possible to make artificial skin for patients with burns or skin ulcers. Clinical trials are under way to make even corneas, cartilage, bone and liver with polymer/cell combinations.

A stent is an artificial tube that prevents an artery, the urinary tract or other passage getting blocked. A stent can now be made from a shape memory material, which changes shape from a flexible string at room temperature to a firm tube at body temperature. Ballooning outwards it opens up the blocked passage.

e Suggest an advantage of using artificial skin rather than having a skin graft.

f How do shape memory stents prevent patients needing invasive surgery?

Practical

Two-way memory

1 Bend the smart wire and transfer it to a beaker of hot water (at about 90°C).

2 Transfer the two-way smart spring to hot water and observe.

3 Place the smart polymer into hot water with tongs and twist to a new shape. Allow to cool and then return it to the hot water.

● How do the two-way memory materials behave at different temperatures?

● How do the atoms of a smart alloy rearrange at different temperatures?

Safety: Take care using hot water. Avoid smart alloys if you have a nickel allergy. Wear eye protection.

Summary questions

1 Name a product that has evolved over time and describe the properties of the materials used in the design nowadays.

2 Why is an X-ray shielding tungsten–polymer composite better than solid lead and tungsten for the clothing of a radiographer?

3 Select three different examples from the table below, from different categories (sports, medicine and transport). Explain why each material is used based on the properties of the materials.

Types	Materials	For sports	For medicine	For transport
Metals	Aluminium with silicon or titanium alloy	Tennis rackets	Replacement hip joints	Aircraft structures Bicycle frames
	Stainless steel (iron with carbon, chromium, nickel, molybdenum)	Golf clubs	Surgical instruments/ implants MRI scanners	Ships tankers Car bodies/ exhausts
Polymers	Lycra	Body suits	Support stockings	Car bumpers
	Polypropene (PP)	Floating rope	Sutures, valves, joints	Expanded for model aircraft
Ceramics	Ceramics mixtures Carbon-fibre ceramic	Clay pigeon shooting Racing car brakes	Dentures	Space shuttle shell tiles Catalytic converters
Composites	Titanium with carbon or Kevlar fibres in resin	Tennis rackets Golf clubs	Prosthetics	Bicycle frames

Key points

● Properties such as insulation, density, strength, smoothness, flexibility and shock absorbency affect the suitability of a material for a product.

● Using synthetic polymers and composite materials has influenced design. People are no longer reliant on natural materials and pure metals.

3.8 Standard procedures for testing materials ⓚ

Testing materials

Scientists must first test materials before deciding on their suitability for different applications. They perform standard tests on materials to measure their properties. Some **standard procedures** that materials scientists use are explained here and others are revised.

🖩 Maths skills

Density

Calculate the density of the material using:

$$\text{density} = \frac{\text{mass}}{\text{volume}} \text{ (in g/cm}^3\text{)}$$

Measure the object's mass (in g).

For a regular-shaped object, use a formula to calculate the volume.

For a cuboid the volume (in cm^3) = length (cm) × width (cm) × height (cm)

For an irregular-shaped object, find the volume by displacing water in a measuring cylinder.

The volume of the object = the volume of water displaced (in cm^3)

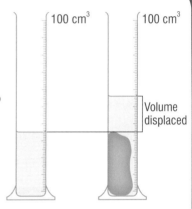

Figure 1 Measuring the density of an irregular-shaped object

Practical

Hardness test

Figure 2 The hardness test

Select some metal, wood and polymer samples.

Place a petroleum jelly-coated ball-bearing on your sample of material.

Drop the 1 kg mass, from the same height each time, onto the ball-bearing carefully.

Measure the diameter of the dent made in the sample with calipers. Repeat with another sample.

Safety: Avoid dropping the falling masses onto anyone's fingers and feet. Protect bench tops.

Practical

Tensile breaking strength

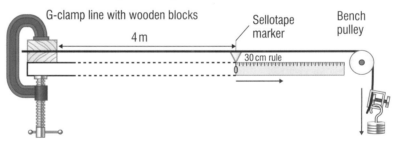

Figure 3 Tensile (tension) test apparatus

Add a Hoffman clip to straighten the line (it acts as a useful hanger for the weights). Then add a Sellotape marker at the zero mark.

Compare cotton and polyester thread with solid (mono) and stranded nylon (fishing) line. See how each stretches as you add loads. Note the tensile breaking strength (in N).

Measure the radius of each line with a micrometer in millimetres. Calculate the area of each line (in mm^2) using: $A = \pi r^2$.

Calculate the tensile breaking stress of each thread (in N/mm^2) using:

$$\text{stress} = \frac{\text{force (N)}}{\text{area (mm}^2\text{)}}$$

Safety: Avoid falling masses.

Practical

Compressive breaking strength

Fit a thin plastic rod into the pre-drilled coins in the syringe.

Add masses to the bar until the specimen buckles or breaks. Remember a mass of 1 kg has a weight of 10 N. As the bar acts as a lever, the breaking force is ten times the load added.

Measure the radius of the rod in millimetres.

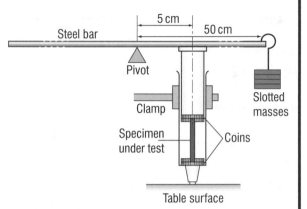

Figure 4 Loaded cantilever method

Calculate the area of cross-section of the rod (in cm²) using: $A = \pi r^2$.

Calculate the tensile breaking stress of each rod (in N/mm²) using:

$$\text{stress} = \frac{\text{force (N)}}{\text{area (mm}^2)}$$

Safety: Avoid falling masses. Wear eye protection.

Practical

Flexibility/stiffness test

Figure 5 The flexibility test

Select metal, wooden and polymer rods with the same dimensions. Place a 100 g mass on the end and measure the deflection.
Repeat with masses up to 1 kg.
Repeat with another material.

Safety: Avoid falling masses. Protect the floor.

Remembering material tests

Use the following table to help with revision. Remember to carry out a risk assessment before testing.

Students need to know:	Topic	Further guidance
Resistance to corrosion	Making connections	Set up test tubes to show which metals are resistant to corrosion and what conditions cause the most corrosion.
Electrical conductivity	3.1	Measure the resistance with a multimeter. A poor conductor resists.
Toughness/brittleness	3.4	Measure the hammer angle to fracture the specimen. Observe the way the materials fracture.
Thermal conductivity	3.5	Compare the times for heat to pass through different materials.
Hardness	1.5/3.8	Measure dent size when a mass falls onto the material.
Stiffness/flexibility	3.8	Measure the deflection of the end of a loaded rod.
Density (= mass ÷ volume)	3.1/3.8	Obtain volume using a formula or by displacement. Obtain mass in grams and calculate density.
Compressive breaking strength	3.8	Use the loaded cantilever method.
Tensile breaking strength	3.2/3.8	Note the force breaking the wire. Calculate breaking stress = force ÷ area.

Summary questions

1 What is the density of an object of mass 48 g that removes 12 cm³ of water when it is placed in a displacement can?

2 Define 'hard'.

3 What is the connection between flexibility and stiffness?

4 A thread of diameter 2 mm breaks when you apply a force of 100 N. What is:
   a its cross-sectional area in cm²
   b its tensile breaking stress in N/cm²?

Key points

- Use the units of density (g/cm³) for calculated values.

- A hard material does not dent or scratch.

- A stiff material is not flexible.

- Add the units N/mm² to stress calculations.

- Revise all the techniques listed in this topic, including those referred to in the maths skills box.

Summary questions

1. Copy and complete the sentences using the words below:

   Standardisation   Standards   purpose

   Scientists test materials to assess their fitness for .... The British .... Institute and European Committee for .... set and test product standards.

2. **a** Calculate the density of a 45 g bar of titanium with dimensions 5 cm, 2 cm and 1 cm.
   b Give a sports, medical and transport use for titanium, saying why its low density is of benefit to the product.

3. Copy and complete the sentences using the words below.

   Hooke's   constant   compression

   Both tension and .... affect the shape of objects. N/cm is the unit for the spring .... , describing how a material stretches when it obeys .... law.

4. **a** What is Hooke's law?
   b An elastic spring with constant 0.5 N/cm stretches by 6 cm. What force acts on the spring?
   c Explain which spring has a bigger spring constant, one for a car suspension or one for a trampoline.

5. A tungsten wire (of diameter 0.4 mm) breaks when stretched with a force of 200 N.
   a What is the radius of the wire?
   b Use $A = \pi r^2$ to calculate the area of the wire.
   c Show that its tensile breaking strength is approximately 1600 N/mm².
   d Car light bulb filaments are made from tungsten. Suggest why its strength provides a long bulb-life even in off-road vehicles.

6. Copy and complete the sentences using the words below:

   *aluminium   artificial   bicycle   energy   fibre
   low   replacement   resistant   shipping
   strong   surgical*

   Aircraft frames are made from .... density .... alloy. Stainless steel is used for .... instruments and .... containers. Lightweight, ... titanium is used in .... hip joints. Kevlar composites absorb .... so are useful for tennis rackets. Carbon .... composites are found in .... frames. The Space Shuttle has heat-............. ceramic tiles. Polymer-cell combinations are used for .... skin.

7. Copy and complete the sentences using the words below:

   *Composites   conduct   density   hammering
   hard   high   low   malleable   mass
   properties   strength   strong   thermal*

   Polymers are flexible with low .... and poor .... conductivity. Metals are .... as they can be shaped by .... or rolling. Metals are usually ...., hard and .... heat well. Ceramics are .... and brittle with .... melting points. .... are made of more than one material, making the most of the .... of each. Carbon fibre/epoxy resin composites have .... density and high .... Density is measured by dividing .... by volume.

8. Copy and complete the sentences using the words below:

   *area   density   flexible   natural
   prevent   run   shock   synthetic*

   Cotton and leather are .... materials. Polyester and Lycra are .... or man-made materials. Sports clothing needs to be lightweight and .... Insulation helps .... heat loss. A large surface .... loses heat more quickly than a small one. The lining of trainers is often .... absorbing. The overall .... of sports shoes is low, allowing the athlete to .... more quickly.

9. **a** What happens to a thermoplastic polymer when it is heated? **[H]**
   b What benefit is this in the manufacture of vending cups? **[H]**

10. Explain why thermosetting polymers cannot be remoulded. **[H]**

11. **a** How does increasing the chain length and amount of cross-linking affect the properties of a polymer? **[H]**
    b Why do its properties change in this way? **[H]**

Key practical questions

1 Describe the following standard procedures to compare and measure materials with different:
   a resistance to corrosion
   b thermal conductivity
   c hardness
   d stiffness/flexibility
   e density
   f toughness/brittleness
   g compressive breaking strength
   h tensile breaking strength
   i electrical conductivity.

AQA Examination-style questions

1 Materials scientists try and improve sports equipment to help maximise performance.

a Tennis rackets have changed a lot in the last 50 years. Until 1965, professional tennis rackets were made of wood. After 1965, the tennis rackets were made out of metal.

  i Give **one** disadvantage of a wooden tennis racket. *(1)*

  ii Give **two** properties of metals that make them suitable for tennis rackets. *(2)*

b What is the name given to the ability of the racket strings to resist stretching when hit by a tennis ball? *(1)*

c i Racket strings used to be made out of animal gut. Now they are more likely to be made from a synthetic polymer. Give **three** advantages of using synthetic materials rather than natural materials for making strings. *(3)*

  ii Describe how the bonding of a polymer must be changed to increase its strength. **[H]** *(2)*

  iii Why are thermoplastic polymers not suitable for tennis strings. **[H]** *(1)*

d A tennis player chooses the shoes he wears to play in very carefully. Suggest **two** properties a tennis player would look for in a suitable pair of tennis shoes. *(2)*

2 Kevlar is a modern material which is used in many applications.

a What type of material is Kevlar? *(1)*

b Give **three** properties of Kevlar that make it a good material for making a tennis racket. *(3)*

c Kevlar is also used in vests for police officers. Explain, giving **two** reasons, why Kevlar is a good material to use to make the police vests. *(2)*

3 Material scientists often use metals because of their range of useful properties.

a Which of these is a property of metals? Choose your answer

  conductor   insulator   brittle   soft *(1)*

b Metals are used as ballast to weigh a ship down. This means the ship will not flip over when on the water. A materials scientist is testing iron or lead to see which is most suitable for use as ballast in a sailing ship. He measures the weight and the volume of each of the metals and then works out the density of each metal.

	Mass (kg)	Volume (m³)	Density (kg/m³)
iron	35000	5	7000
lead	22680	2	

  i Using the equation below, find the density of lead. *(1)*

$$\text{Density} = \frac{\text{mass}}{\text{volume}}$$

  ii Using the information in the table explain which metal, iron or lead, the material scientist should use as ballast. *(3)*

4 The diagram shows a golf club and details of the covering of the handle.

  Handle

  Shaft

  Head

a i Suggest a type of material from which the covering of the handle could be made. *(1)*

  ii Suggest why the covering of the handle has an uneven (dimpled) surface. *(1)*

b The table shows some properties of materials used in the manufacture of golf clubs.

Property	Titanium	Aluminium	Steel	Carbon fibre-reinforced plastic
Density (g/cm³)	4.5	2.7	7.7	2.1
Strength (GPa)	120	70	200	150
Melting point (°C)	1668	660	1370	250
Corrodes	No	No	Yes	No
Hardness (10 = hard)	6	2.7	4.2	3.5

Use the information in the table to answer the following questions.

  i Give **two** advantages of using carbon fibre reinforced plastic for the shaft of the golf clubs. *(2)*

  ii The heads of some golf clubs are made of titanium. Give **two** reasons why this metal is used. *(2)*

c What type of material is carbon fibre reinforced plastic? *(1)*

d Golf shoes can be made from leather, which is a natural material, or from a synthetic material. Give **two** advantages of making the golf shoes out of a synthetic material rather than leather. *(2)*

AQA, 2010

1 Bicycles are one of the world's most popular modes of transportation.

> Before the 1970s bikes were constructed of heavy, but strong, steel and alloy steel. Manufacturers wanted to use new materials to increase strength, rigidity, lightness, and durability. In the 1970s more versatile alloy steels were used which were lighter and more inexpensive. In the 1980s lightweight aluminum frames became the popular choice. The strongest metals, however, are steel and titanium with life-expectancy of more than one decade, while aluminum may only last three to five years. In the 1990s manufacturers used even lighter and stronger frames made of composites of structural fibres such as carbon.

a Give **two** desired properties of a material used to make a racing bike? *(2)*

b What is meant by the term composite? *(1)*

c The table shows some properties of three materials.

	Density (kg/m³)	Melting point (°C)
aluminium	2712	660
steel	7500	1535
titanium	4500	1668

   i Use the information in the table to choose the best material to make a bike frame. Give a reason for your choice. *(2)*

   ii Give a disadvantage of using aluminium rather than steel or titanium to make a bike frame. *(1)*

d Give **two** ways the materials used for making bike frames have improved since the 1960s. *(3)*

AQA *Examiner's tip*

If the examiner has given a paragraph to read then it is important that you spend some time reading it. For this question a lot of the information to help you with the answers can be found in the paragraph.

2 Materials scientists need to know the properties of the material they want to use.

a Match the material on the left to its correct property on the right.

Materials	Properties
Aluminium	is strongly magnetic
Ceramic	lightweight
Kevlar	a good heat insulator
Polystyrene	resistant to high temperature
	absorbs energy

                                                              (4)

b Titanium alloys are used to make aircraft frames. What is the meaning of the term alloy? *(1)*

c A student made a bridge out of plywood. He tested the bridge to see how much weight it would hold before it broke. Below is a diagram of his experiment.

Load—⊙   Plywood

   i What type of material is plywood? *(1)*

   ii What force is acting at the top of the plywood? *(1)*

   iii What force is acting at the underside of the plywood? *(1)*

d The plywood bridge began to bend when a load of 6 Newtons was placed on the bridge. The plywood has a cross sectional area of 10cm². Calculate the stress the load put on the bridge to make it bend. Give the units. *(2)*

$$\text{Stress} = \frac{\text{force}}{\text{cross-sectional area}}$$

AQA *Examiner's tip*

This is a calculation question. Remember to show your working. The question also asks for the units. This means there will be a mark for the correct units so make sure you get this right.

3 A new hospital has been opened near your local town.

  a The doctors and nurses wear clothing called scrubs when they are in the operating room. A materials scientist tested different materials to see which would be best for making the scrubs. Her results are shown in the table below.

Property	Polypropylene	Polyester	Wool	Cotton
Absorbancy	Very low	Low	High	Medium
Colour fastness	Very good. Pre-coloured in manufacture	Good	May fade in washing	May fade in washing
Durability	Excellent	Excellent	Fair	Fair
Stain resistance	Resists stains	Requires stain-release chemicals	Requires dry cleaning for stain release	Requires bleaching for stain removal

  Which fibre would you use to make the scrubs? Use the information in the table to explain your choice. *(4)*

  b All the surgical instruments in the hospital have a mark on them showing that they have been tested for their quality and fitness for purpose in Europe. Give the name of this mark? *(1)*

  c *In this question you will be assessed on using good English, organising information clearly and using specialist terms where appropriate.*

  A materials scientist is making surgical instruments to use in an operating room. He cannot decide whether to use stainless steel or titanium alloys to make the instruments.

  Explain what properties he would need the surgical instruments to have. Describe how the materials scientist would test each property to determine which material was most suitable for the surgical instruments. *(6)*

4 a Hockey shirts can be made from natural or synthetic fibres. Copy and complete the table below using ticks to show whether the material is a natural or a synthetic fibre.

Fibre	Natural	Synthetic
cotton		
nylon		
polyester		
wool		

  b Hockey sticks used to be made from wood but are now made from carbon fibre. Give **two** properties that make carbon fibre a good material for making hockey sticks. *(2)*

  c Figure 1 shows a helmet used by a hockey goalkeeper.

  The table show the properties of some material that could be used to make a faceguard.

	Hardness	Density (kg/m³)	Durability
aluminium	low	very low	high
steel	high	high	medium
titanium	high	low	high

  Which metal would you choose to make the faceguard?
  Use the information in the table to explain your choice. *(3)*

  AQA, 2008

Figure 1

AQA **Examiner's tip**

It is important that you give a property that is linked to the use (in this case as a faceguard). You will not get a mark if you choose a property of the metal but it is not related to the use.

Making connections

In this chapter you will learn how food scientists monitor the quality and safety of the food you eat. A 'food scientist' is the general name given to any scientist who works with food. An example of a food scientist is a microbiologist. One role of a microbiologist is to detect the presence of unwanted organisms in our food, which may make us ill. They also advise food producers, such as bakers and brewers, on the optimum conditions for working with useful microorganisms.

You will also study how agricultural scientists have helped to maximise food production. Agricultural scientists work to develop chemicals such as fertilisers, and investigate the use of controlled environments to maximise food yields.

Figure 1 Food scientists analyse foods to determine the levels of different nutrients accurately. This ensures you know what is in the food you are eating, so you can make an informed choice about what to buy.

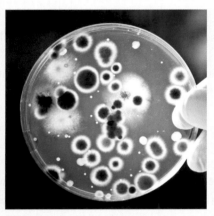

Figure 2 Microbiologists detect the presence of harmful microorganisms in food. If undetected, these can cause food poisoning.

Figure 3 Dieticians advise people about the foods they should eat and those they should avoid. These include people with medical conditions, such as diabetes and allergies.

Figure 4 Microbiologists determine the optimum conditions for microorganisms to grow. This helps to improve the production of products like beer, bread and yoghurt.

Figure 5 Agricultural scientists have developed chemicals to increase farming yields. These chemicals include fertilisers, pesticides, fungicides and herbicides.

Figure 6 Food scientists have genetically engineered some food products, such as frost-resistant tomatoes. Techniques like this are used to help feed the world's increasing population.

a Name three products made using microorganisms.

b What do dieticians do?

Microbiologists

A 'microbiologist' is the name given to a food scientist who works with microorganisms. Microbiologists perform a number of different roles. These include:

- Public health laboratory – microbiologists test products for the presence of unwanted microorganisms such as *E.Coli*. The presence of these organisms in food can cause food poisoning.
- Yoghurt production – microbiologists work with different combinations of useful bacteria to produce an ideal blend for making yoghurt. They also determine the optimum conditions the bacteria should be kept in to produce high yields.
- Manufacturing insulin – microbiologists work alongside other scientists to genetically engineer bacteria, to produce useful chemicals such as insulin.

c What is the name of a scientist who works with microorganisms?

d Name a drug which bacteria can be genetically engineered to produce.

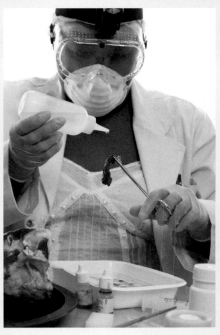

Figure 7 A microbiologist inspecting a food sample

Activity

The role of microbiologists

Produce an A4 leaflet describing the specific role of one type of microbiologist. You can choose one detailed above, or carry out some research to find another industry in which microbiologists are employed. Your leaflet should state the role of the microbiologist, the industry that they work in, and the importance of their work in the food industry.

Scientists @ work

Agricultural scientists

Agricultural scientists perform a wide range of roles including developing new crops, managing the environment and producing a greater yield of plants and animals from an area of land.

- Carry out an internet search to find out about the specific roles carried out by agricultural scientists. Try to come up with 10 different roles. As a whole class, discuss the wide range of jobs they perform. How has their knowledge, and the technology they have developed, increased food production throughout the developed world?

4.1 Introduction to food science

The Food Standards Agency

The Food Standards Agency (FSA) was set up in 2000 by an Act of Parliament. It is an independent authority which protects the public's health and consumer interests relating to food. **Food scientists** work within the FSA to promote good eating habits and the importance of a healthy diet. They also make sure that your food is safe, and that it is labelled correctly.

Food scientists and **dieticians** think that everyone should eat a healthy diet containing lots of fruit and vegetables. Most energy should come from starchy foods such as pasta and rice. You should also have a low intake of fat (especially saturated fat), salt and sugar.

a What does FSA stand for?

Department for Environment, Farming and Rural affairs

Farming in the UK is monitored by the Department for Environment, Farming and Rural Affairs (**Defra**). They make certain that farmers produce a food supply which is healthy and sustainable. This in turn ensures that the farming industry is thriving. **Agricultural scientists** work within Defra to study farm crops and animals. They develop new techniques to improve the quality and quantity of farm products.

b What does Defra stand for?

Food scientists

Dieticians, public health inspectors and food analysts are all examples of food scientists. Food scientists use a number of standard procedures to perform their role, including:

- analysing the quality of food products – to ensure that a product contains what the manufacturer claims on labelling and advertising
- checking the safety of food products – to check for the presence of microorganisms, and determine the type and number present
- applying scientific techniques to keep food fresh, safe and attractive
- using chemical tests to detect the presence (and quantity) of vitamins, fat, sugar, protein and additives in a food product
- producing new food products, or producing food more quickly and cheaply
- developing food supplements, and studying their effect on health.

c State two key roles a food scientist performs.

d Name three substances that a food scientist can test in a food product.

Figure 1 This food scientist works in quality control analysis, checking the nutritional value of a food product

⊂⊃ links

For more information on how food scientists use standard procedures to test foods see 4.16 Standard procedures used in food science.

Scientists @ work

Where are food scientists employed?

Food scientists are employed in lots of different ways. Dieticians are a particular type of food scientist. They help people to choose a healthy diet.

Microbiologists investigate outbreaks of food poisoning such as *Salmonella*.

Public health laboratories

Dieticians study people's diets. They look at what a person eats, and suggest ways to help make their diet healthier. In hospitals dieticians prepare special diets for people who suffer from allergies or have conditions like diabetes or some forms of cancer.

Dieticians work with lots of different groups of people to promote healthy eating. For example, they could advise a children's nursery on the meals they should be providing for a particular age group.

Hospital

Community

Where are food scientists employed?

Media

Agriculture

Dieticians make sure that the public are given good advice about healthy eating. For example, the government's 'Five-a-day' campaign.

Agricultural scientists help to develop improved crops, for example, genetically engineered, drought-resistant wheat. Other roles of an agricultural scientist include:

- studying how to control crop pests and weeds effectively
- ensuring that soil and water sources are conserved
- investigating how to produce food products more efficiently.

Sports training villages

Nutritionists help athletes understand the links between their performance and diet.

Food manufacturers and food retailers

Food analysts make sure that food being made and sold contains what it says on the label.

Figure 2 This dietician is advising a man about which foods to avoid to lose weight and which foods are rich in nutrients that he needs to ensure a healthy diet

Figure 3 This microbiologist is removing a sample from a defrosted chicken to check for bacteria

Key points

- The FSA makes sure that the food we eat is safe, labelled correctly and helps to promote healthy eating.

- Defra ensures farmers produce healthy and sustainable food.

- Food scientists use a range of standard procedures. These include monitoring nutrient levels, and checking for the presence of unwanted microorganisms in food products.

- Food scientists include nutritionists, dieticians, food analysts, agricultural scientists and public health inspectors.

e Name three conditions a dietician could help to treat or manage.

Summary questions

1 What are the three main roles of the Food Standards Agency?

2 Name five areas in which food scientists are employed. For each example, describe the type of work which is carried out.

3 Explain why the work of food scientists is important.

4.2 Food poisoning

Learning objectives

- What causes food poisoning?
- What are the optimum conditions for bacterial growth?
- What are the symptoms of food poisoning?

 Did you know ... ?

There are more living organisms on the skin of a single human being than there are human beings on the surface of the Earth! Make sure you wash your hands thoroughly before cooking food.

Food poisoning is caused by the growth of microorganisms in food. The most serious types of food poisoning are due to bacteria, and the toxins they produce. Microbiologists and public health inspectors are responsible for monitoring the growth of bacteria in places where they could cause a harmful effect. These include restaurants and food manufacturing companies.

What types of bacterium cause food poisoning?

There are three main groups of bacterium that cause food poisoning:

- *Campylobacter* – this is the most common cause of food poisoning. *Campylobacter* can be found in raw meat, unpasteurised milk, and untreated water.
- *Salmonella* – found in a wide range of food products including raw meat, eggs, raw unwashed vegetables and unpasteurised milk.
- *E. coli* – there are many different types of this organism; some are beneficial, others are capable of causing illness. *E. coli* 0157 can cause severe illness, and is found in raw and undercooked meats, unpasteurised milk and dairy products.

All these bacteria can survive refrigeration and freezer storage, but thorough cooking of food, or pasteurisation, will kill them.

a Name the three main groups of bacterium responsible for food poisoning.

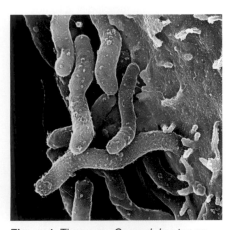

Figure 1 These are *Campylobacter* on the surface of the human stomach

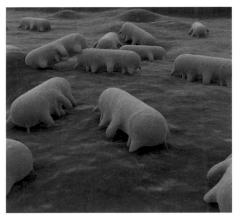

Figure 2 These are *Salmonella* bacteria

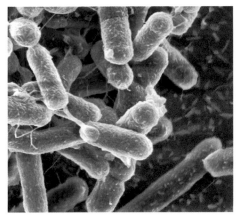

Figure 3 This is *E. coli* 0157.

What are the optimum conditions for bacterial growth?

Bacteria multiply rapidly – in the right conditions one bacterium can multiply to more than 4 million in just 8 hours.

To grow and reproduce effectively most bacteria need three things:

- warmth
- moisture
- a food source.

These conditions are present in most fresh foods. This is why fresh food can only be used for a short period of time before the food 'goes off'.

b What conditions do bacteria need to multiply?

What are the symptoms of food poisoning?

Common signs of food poisoning are stomach pains, diarrhoea, vomiting and fever. These symptoms usually appear rapidly, but can occur several days after eating contaminated food. Most people get better within a few days.

Over-the-counter medicines, such as oral rehydration salts, can be taken to replace fluid and salts lost through diarrhoea. In more serious cases, fluids may need to be replaced through a saline drip. In rare circumstances, food poisoning can kill.

c How would you feel if you were suffering from food poisoning?

 Did you know ...?

Each year it is estimated that as many as 5.5 million people in the UK suffer from food-borne illnesses – that's 1 in 10 people!

Activity

Food poisoning

When food products become contaminated, it can result in the products being recalled or health scares. For example, contamination with *Salmonella* led Cadbury to recall over a million bars of chocolate.

Carry out some research into an outbreak of food poisoning, and then discuss the following questions:

- Who is most at risk from food poisoning outbreaks?
- Why do companies recall products which may be contaminated?
- What problems does this cause for the company?
- What is the response of the media to outbreaks of food poisoning?
- What is the role of public health inspectors, and the Health Protection Agency?

Summary questions

1 Copy and complete the sentences using the words below:

vomiting   bacteria   Salmonella   poisoning   diarrhoea   drinks

Food .... is normally caused by bacteria present in foods and ....

.... and *E.coli* are examples of .... that can cause disease.

People who suffer from food poisoning often suffer from .... and ....

2 How can oral rehydration salts help someone suffering from food poisoning?

3 How can the following conditions prevent bacteria growing in food?
a Freezing
b Drying

Key points

- Food poisoning is caused by the growth of microorganisms (usually bacteria) in food.

- For optimum growth, bacteria need warmth, moisture and a food source.

- Common signs of food poisoning are stomach pains, diarrhoea, vomiting and fever.

4.3 Food hygiene

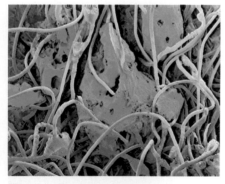

Figure 1 These are food particles trapped in a cleaning cloth. They provide an excellent breeding ground for bacteria. Hands, work surfaces and utensils must be cleaned regularly to prevent the cross-contamination of clean areas.

Public health inspectors are responsible for ensuring premises which produce, store or handle food products are kept clean. These include kitchens, restaurants and food manufacturers. These establishments must be kept spotlessly clean to ensure food products are not contaminated with microorganisms. This reduces the risk of food poisoning.

What procedures can be followed to prevent food poisoning?

Microorganisms are very hard to detect, as they do not usually affect the taste, appearance or smell of food. Microorganisms can enter our food at any time. This could be from when an animal or crop is growing, until the food is eaten. If they survive and multiply, they can cause illness.

a Why is it hard to detect the presence of microorganisms in food?

There are a number of procedures which are used to ensure that foods are prepared with as little risk of contamination as possible. They include:

- using detergents to ensure food preparation areas remain clean
- using disinfectants to kill bacteria on work surfaces
- using sterile packaging materials
- sterilisation of equipment using high temperatures or gamma rays
- disposing of waste into appropriate containers, and regularly removing waste from food preparation areas
- adequate control of pests, such as insects or mice.

b Why should you disinfect kitchen worktops regularly?

What is meant by 'good personal hygiene'?

People who work in the food industry must maintain good personal hygiene. This helps to avoid potentially harmful microorganisms (pathogens) from being transferred to foods. Examples of good personal hygiene include:

- washing hands with soap – before meals, before preparing food and after going to the toilet
- taking a bath or shower daily – especially in hot weather, as microorganisms thrive in these conditions
- tying hair back when working with food, or wearing a hair net
- covering cuts with a plaster – if working in the food industry these should be blue, so they can easily be seen if they fall off in the food!
- wearing protective clothing – for example disposable gloves and an apron.

c Name three examples of protective clothing a person should wear when preparing food.

How are bacteria prevented from growing in food products?

The growth of bacteria can be slowed down or stopped altogether by changing the temperature. Bacteria multiply rapidly between about 5°C and 65°C. Most are killed at temperatures above 70°C.

Technique to preserve food	How most bacteria are prevented from growing	Example of food product prepared in this way
Refrigeration	Slows down, but does not stop the growth of bacteria. If the product is warmed up, the bacteria will begin to multiply	Fruit juice
Freezing	Stops bacteria multiplying but does not kill them	Ice cream
Heating – for example, ultra-heat treatment (UHT) heats food to 132°C for one minute	Kills virtually all microorganisms and their spores	Milk
Cooking	At the correct temperature, kills microorganisms	Chicken
Drying	Removes water, so bacteria cannot digest and absorb the food source	Rice
Salting	Bacteria lose water from their cells by osmosis. Most die, others are inactivated (can't reproduce)	Corned beef
Pickling	Acid (usually vinegar) is added to lower the pH of the food stuff. This inactivates most microorganisms	Pickled onions

d Why does cooking chicken thoroughly prevent you suffering from food poisoning?

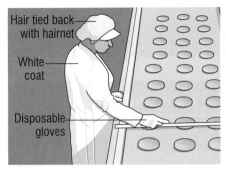
Figure 2 People working with food should maintain good personal hygiene and wear protective clothing

?? ? *Did you know ...?*

The term 'UHT milk' refers to ultra-heat treatment. The milk is heated to 132°C for one minute and then rapidly cooled. This kills virtually all microorganisms, and their spores. Milk produced in this way can be stored for longer periods than pasteurised milk.

Summary questions

1 Copy and complete the sentences using the words below:

microorganisms   cooking   contamination
food poisoning   hygiene

.... is caused by .... growing in food. Good .... prevents .... of food, and thorough .... kills any microorganisms present.

2 Why does food keep for longer when stored in a fridge?

3 Why does vacuum-packing, or placing food in an airtight container, help to prevent food poisoning?

4 Produce a list of good practices that could be used in a restaurant kitchen to ensure that food is produced as safely as possible.

Key points

● Food poisoning is avoided in commercial food preparation areas by ensuring work areas are clean, using sterile packaging, and by effectively controlling pests.

● Good personal hygiene is essential for employees in the food industry. This is achieved through: regularly washing hands, covering cuts, and wearing appropriate protective clothing.

● Bacterial growth in food can be inhibited in a number of ways. These include: freezing, ultra-heat treatment, thorough cooking, pickling and drying.

4.4

Standard procedures – microbiological techniques 🄺

Learning objectives

- What is aseptic technique?
- How can you identify the species of bacteria present in a sample?
- How can you accurately count the number of bacteria in a sample?

Food scientists and microbiologists are responsible for checking for the growth of bacteria in food and drink products. Sometimes scientists want to improve the growing conditions, so that bacteria replicate as quickly as possible. An example is in the production of yoghurt.

More commonly scientists want to prevent the growth of unwanted bacteria, as they would spoil food and could cause food poisoning. Techniques commonly used by microbiologists are as follows.

Aseptic technique

Aseptic means 'without microorganisms'. Scientists use aseptic technique to prevent unwanted microorganisms from entering a sample they are studying. It also stops microorganisms from passing from the sample on to themselves, possibly causing disease.

Microorganisms are often transferred from one medium to another with a wire loop. Before the loop is used it must be sterilised. To sterilise the loop, it should be heated until it glows red in a Bunsen burner flame. Then allow the loop to cool before use. While cooling, the loop should be held close to the flame, away from the bench, to ensure it remains sterile.

   a How do you sterilise a wire loop?

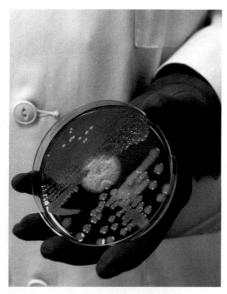

Figure 1 This laboratory technician is holding an agar plate containing several types of bacterium (yellow/green) and a fungus (white) cultured from an environmental swab

Sampling the environment

Bacteria are commonly found throughout the environment. A sterile, cotton swab is used to sample bacteria on a surface such as a desktop, the floor, or a stair handrail. The standard procedure below should be followed:

1 Wipe the swab across the surface being sampled.
2 Tilt lid and rub the swab lightly on the surface of an agar plate.
3 Put the lid on and secure by four small pieces of tape. ***Do not seal all the way round the lid.***
4 Incubate at 25 °C for around 48 hours to allow any microorganisms present to grow.
5 Identify any microorganisms present.

   b Why does the agar plate need to be incubated for a few days?

Making a streak plate

This technique is used to isolate individual bacterial colonies, so that they can be identified. Bacteria can be recognised by the colony they form – they differ in characteristics like shape, colour, size and elevation. The following standard procedure should be used:

1 Dip a sterilised wire loop into the sample of bacteria.
2 Tilt lid and make four or five streaks across one side of an agar plate.
3 Flame and cool the loop.
4 Make a second series of streaks by crossing over the first set, picking up some of the cells and spreading them out across a new section of the plate.

⁇ Did you know ... ?

Anaerobic bacteria (bacteria that reproduce in a lack of oxygen) are generally more dangerous to our health than bacteria grown on agar plates with a good supply of oxygen. This is why agar plates must not be sealed all the way round.

5 Repeat steps 3 and 4 two more times making a third and fourth set of streaks. Fix lid with four short lengths of tape. **Do not seal all the way round.**

6 Incubate the plate upside-down allowing the cells to form colonies. Do not open the plate. Dispose of plates in disinfectant or sterilise.

Serial dilutions to accurately count bacteria

It is extremely difficult to count the number of bacteria in a food sample using a microscope, because there are so many bacteria. Diluting a bacterial sample spreads out the bacteria, so individual colonies of bacteria can be seen on an agar plate. These can be seen as individual dots on the plate. By counting the number of colonies (dots) that grow, the number of bacteria present in the original sample can be calculated:

$$\text{Number of bacteria (per cm}^3\text{) in original sample} = \text{number of colonies} \times \text{dilution of sample}$$

If more than one type of colony is visible, there is more than one species of bacteria present in the sample. You should note how many of each type of colony are present. Then use the formula above to determine how many of each type of bacteria are present.

> c Why do scientists not count individual bacteria?

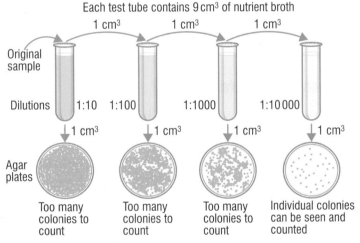

Each test tube contains 9 cm³ of nutrient broth

Figure 3 Carrying out a **serial dilution**

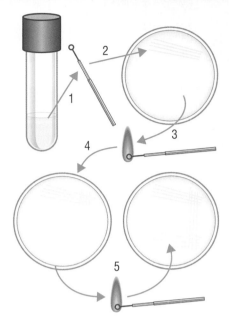

Figure 2 How to make a **streak plate**

Figure 4 This is a streak plate of *E. coli* 0157. This *E. coli* strain was responsible for an outbreak of food poisoning in Scotland in 1996. Over a dozen people died and hundreds were taken ill.

Key points

● Microbiologists work in aseptic (sterile) conditions to ensure samples they are investigating are not contaminated. This also ensures harmful bacteria do not cause the scientists to become ill.

● Microbiologists isolate individual colonies of bacteria using streak plates. The colonies are then identified by looking at their characteristics.

● The number of bacteria present in a sample can be calculated by carrying out a serial dilution.

Summary questions

1 What safety precautions should be taken when working with microorganisms?

2 Describe a situation where scientists from the Food Standards Agency might want to check foods for the presence of bacteria.

3 Why must agar plates not be fully sealed?

4 A 1 cm³ bacterial sample was diluted to 1/100 000th of its original concentration, to allow counting of colonies. If 46 colonies were counted, how many bacteria were present in the original sample?

4.5 Bread, beer and wine production

Bacteria and fungi play an important role in the production of some foods and drinks. Microbiologists study these microorganisms to find out the optimum conditions for their growth – warmth, moisture and a food source. Using this knowledge, scientists can provide the best conditions for microorganisms to grow. This has resulted in many useful food products being made.

Yeast is a type of fungus. It is needed to make bread, beer and wine. These three products are made using a chemical reaction called **fermentation**.

Fermentation is an example of **anaerobic respiration** – the yeast respires without oxygen to ferment sugar, producing alcohol and carbon dioxide. Fermentation can be summarised in this chemical equation:

$$\text{glucose} \longrightarrow \text{ethanol (alcohol)} + \text{carbon dioxide}$$

$$C_6H_{12}O_6 \longrightarrow 2C_2H_5OH + 2CO_2$$

Enzymes speed up the process of fermentation. These enzymes are found in yeast.

a What is ethanol more commonly known as?

What are the ideal conditions for yeast to work?

Enzymes in yeast speed up the process of fermentation. The ideal conditions for fermentation are a good supply of glucose, with no oxygen present, and at a temperature between 15 °C and 25 °C.

How is bread manufactured?

Bread is made using the fermentation reaction. A baker mixes together flour, water, sugar and yeast to make dough. The dough is then left in a warm place to rise. Enzymes in the yeast change the sugar into ethanol and carbon dioxide gas. This gas is trapped inside the dough, making it rise.

The dough is then baked. In the oven, the ethanol evaporates. The bubbles of gas expand, making the bread rise further.

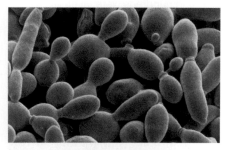

Figure 1 This is *Saccharomyces cerevisiae* – also known as baker's or brewer's yeast

Practical

Making bread

Try making your own bread. Yeast can be bought in the supermarket, in dried form.

- What are the advantages of making your own bread over buying supermarket bread?

b Why is bread not alcoholic?

c Why is dough not put in the fridge to rise?

Figure 2 This baker is kneading bread dough and shaping it into loaves

How is beer produced?

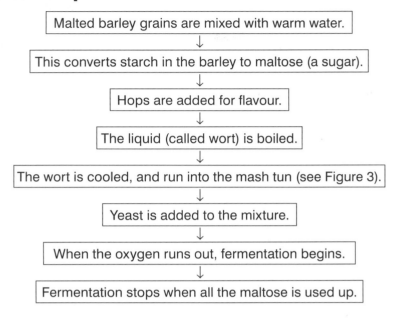

Malted barley grains are mixed with warm water.
↓
This converts starch in the barley to maltose (a sugar).
↓
Hops are added for flavour.
↓
The liquid (called wort) is boiled.
↓
The wort is cooled, and run into the mash tun (see Figure 3).
↓
Yeast is added to the mixture.
↓
When the oxygen runs out, fermentation begins.
↓
Fermentation stops when all the maltose is used up.

d Why does fermentation stop after a period of time?

How is wine produced?

Wine is made in a very similar way to beer. Grapes are crushed and yeast added. Yeast changes the sugar in grapes into alcohol (ethanol). When the fermentation process is complete, the wine is bottled.

∞ links

For information on how food poisoning from pathogenic microorganisms can be prevented, look back to 4.2 Food poisoning and 4.3 Food hygiene.

Figure 3 A 'mash tun', where fermentation occurs in a brewery

---

Practical

Making alcohol

You can make alcohol (ethanol) by adding yeast to fruit juice and sugar. To enable the yeast to ferment the sugars, it needs to be kept warm.

As a class, try to determine the optimum temperature for fermentation to take place.

Summary questions

1 Copy and complete the following sentences using these words:

alcoholic   bread   carbon dioxide   fermentation   fungus

Yeast is a .... It is very important in the production of food and .... drinks. During .... , glucose is converted into ethanol and .... Yeast is involved in making beer and ....

2 Jewish people do not eat yeast at Passover. They still eat bread, but it has to be made without yeast. How would it be different from normal bread?

3 Vodka is made from potatoes. Yeast is involved in fermenting the potato sugars. Write a standard procedure to explain how vodka could be made.

Key points

- Bread is made by mixing flour, water, sugar and yeast. The yeast ferments the sugar. The carbon dioxide given off makes the bread rise.

- Beer is made using hops, barley and yeast. The yeast ferments the sugar (maltose) in the barley, to produce alcohol (ethanol) and carbon dioxide.

- Wine is made from grapes and yeast. The yeast ferments the grape sugars, to produce alcohol (ethanol) and carbon dioxide.

4.6 Cheese and yoghurt production

Learning objectives

- How are microorganisms used to make cheese?
- How are microorganisms used to make yoghurt?
- What happens to milk when it is fermented?

What are the advantages of using microorganisms in food production?

There are a number of reasons why it is advantageous to use microorganisms in food production. These include:

- rapid population growth (they can reproduce very quickly)
- easy to manipulate
- reliable – for example, they are not affected by weather
- economical.

Bacteria are used in the production of both cheese and yoghurt. Microbiologists have studied the growth of these bacteria. They have used this knowledge to develop the ideal conditions for the growth of useful bacteria in yoghurt and cheese manufacture.

How is cheese made?

There are hundreds of types of cheese. Cheese is made from the curdled milk of many animals, including goats, sheep and buffalo. The flow chart below shows simply how cheese is produced.

Did you know ... ?

Cheesemakers can use different bacteria for different cheeses. Some bacteria produce only lactic acid in milk. Others produce lactic acid plus other products, such as carbon dioxide gas. This gas is trapped in the cheese to create the charateristic 'holes' in cheeses such as Emmental.

Milk

Bacteria added to convert lactose (milk sugar) into lactic acid

Rennet added. Rennet contains the enzyme, rennin, which changes a milk protein into casein (curd)

Milk curdles and separates into curds and whey

Whey (mainly water) is drained off

Curds (milk solids – fats and proteins) are pressed to make the cheese solid

Cheese left to ripen to improve its flavour and consistency

Figure 2 This man is adding rennet to milk to make cheese

Figure 1 How cheese is made

a What two substances have to be added to milk to turn it into cheese?

b Why is the whey removed?

Different styles and flavours of cheese can also be made, using different species of bacteria and moulds. Cheeses like Stilton have mould growing throughout the cheese. Others, like camembert, have mould (a type of fungus) on the outer skin.

Fermentation of milk

In cheese and yoghurt production, bacteria are used to ferment sugars found in milk. In both cases, lactose (milk sugar) is converted into lactic acid. It is the acid that gives these products their characteristic tangy taste.

$$\text{lactose (milk sugar)} \xrightarrow{\text{bacteria}} \text{lactic acid}$$

How is yoghurt made?

Yoghurt is another product which is made from milk. Commercially cow's milk is used, but any milk can be turned into yoghurt.

A mixture of bacteria is added to milk which has been pasteurised. This means it has been heated to kill harmful bacteria. The milk is then kept warm for several hours. During this time the bacteria multiply, and ferment lactose into lactic acid. The lactic acid curdles the milk into yoghurt. The acid also hinders the growth of harmful bacteria. This increases the time that yoghurt can be kept and eaten safely.

Yoghurts manufactured from unpasteurised milk are known as 'live' yoghurts.

c What conditions must exist for bacteria to ferment sugar?

Practical

Making yoghurt

Try making your own yoghurt by adding *Streptococcus thermophilus* and/or *Lactobacillus bulgaricus* to boiled milk. Keep the mixture warm overnight in an incubator. Use a food technology area to make the yoghurt if you are going to taste it.

---

d What are the useful properties of lactic acid in yoghurt production?

Summary questions

1 Copy and complete the following sentences using these words:

bacteria   curdles   ferment   lactose   yoghurt

We use .... to .... milk sugar in cheese and .... production. The bacteria change .... into lactic acid. This .... the milk.

2 Why are live yoghurts safe to eat, even though they contain bacteria?

3 Why can you safely store pasteurised foods for a longer period of time than non-pasteurised foods?

Figure 3 Stilton cheese has mould growing throughout it

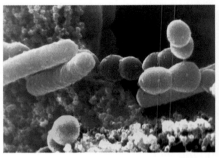

Figure 4 *Streptococcus thermophilus* (red) and *Lactobacillus bulgaricus* (blue) are both used in yoghurt production

Did you know ...?

Pharmacists often recommend that people taking antibiotics eat 'live' yoghurt. This is to replace the useful bacteria which are present in your intestine. These are essential for digestion, but are killed off by the antibiotics at the same time as the pathogenic bacteria.

Key points

● Cheese is made by adding bacteria and rennet to milk. This curdles the milk. The whey is drained off and the curds pressed to make cheese.

● Yoghurt is made by adding bacteria to boiled milk. The milk is kept warm, allowing the bacteria to ferment the lactose (milk sugar). The lactic acid produced curdles the milk, making yoghurt.

4.7 Growing crops ⓚ

Agricultural scientists study how plants grow. They find out which nutrients plants need to maximise their growth. Food scientists have developed efficient techniques for:

- turning raw materials into food products
- increasing the length of time over which food can be safely eaten
- ensuring foods keep many of their nutrients.

There are two different approaches to food production – **intensive farming** and **organic farming**.

Intensive farming

Most of the world's food is produced by intensive farming. Large quantities of food are produced cheaply and efficiently. This occurs by maximising the growth of crops and farm animals. Often this is achieved through the use of chemicals. However, these chemicals can enter food chains, killing animals they were not intended for. Intensive farming techniques are also less labour intensive than more traditional methods. This allows food to be produced more economically.

Organic farming

Organic farmers use natural methods of producing crops and rearing animals. Many people believe food grown in this way is healthier and tastes better. However, yields are generally much smaller. This results in the products being more expensive than those produced by more intensive means.

> **a** State an advantage and disadvantage of intensive farming.
>
> **b** State an advantage and disadvantage of organic farming.

Plant growth

Plants are producers – they make their own food. They convert simple materials found in their environment into glucose (sugar). This process is called **photosynthesis**. They can then use this glucose for growth.

Photosynthesis takes place inside the plant's chloroplasts and is summarised in the equation below:

Word equation: **carbon dioxide** + **water** $\xrightarrow[\text{trapped in chlorophyll}]{\text{light energy}}$ **glucose** + **oxygen**

Symbol equation: $6CO_2$ + $6H_2O$ $\xrightarrow[\text{trapped in chlorophyll}]{}$ $C_6H_{12}O_6$ + $6O_2$

Four factors affect the rate of photosynthesis:

- light intensity
- availability of water
- concentration of carbon dioxide
- temperature.

Figure 1 An industrial-sized greenhouse used to create a controlled environment for plants to grow in

Commercial plant-growers monitor these factors and the rate of photosynthesis is high in their greenhouses by creating a **controlled environment**. For example, by increasing the availability of carbon dioxide, increasing light availability and controlling temperature. The faster photosynthesis occurs, the more glucose is made, and so the more the plant grows.

Adding nutrients to soil

As crops grow, they remove nutrients from the soil. Therefore, to continue to produce good crops from the same land, these nutrients need to be replaced.

● Intensive farmers use artificial **fertilisers** to add nutrients to the soil.

● Organic farmers add nutrients to the soil through natural fertilisers – manure or compost. They also plant leguminous plants like clover, because they add nitrates (an essential nutrient) to the soil.

  c State three ways nitrates can be added to the soil.

Dealing with pests, weeds and fungi

Many insects such as aphids and beetles eat crops. Intensive farmers kill these pests using chemical **pesticides**.

All crop pests have natural predators. Organic farmers can exploit this relationship to kill pests. This is called **biological control**. Predators (normally other insects) are bred in large numbers. They are then released onto crops, where they eat the pests. Parasitic wasps, for example, can be used.

  d Name an example of an organism that can be used to control aphids.

Herbicides are chemicals widely used by intensive farmers. These kill weeds (other plants) that would compete with the crop for water, nutrients and space. On a smaller scale, organic farmers can remove weeds by hand. Machines have been developed to help weed large crop areas without damaging the crop. This method works well on crops that are grown in rows, such as vegetables.

Intensive farmers use chemical **fungicides** to kill fungi that damage crops. Organic farmers rely on growing strong healthy crops to combat disease. However, if a microorganism does attack an organic crop, the farmer will remove the infected material and dispose of it, normally by burning.

Did you know ... ?

Farmers can also use selective breeding and genetic engineering techniques to deal with pests. They can produce new varieties of crops that are more resistant to pests and disease.

∞ links

For more information on how a plant uses specific minerals for healthy growth, see 4.11 Investigating plant growth and 4.8 The use of chemicals in intensive farming.

Figure 2 Ladybirds are used by gardeners and farmers to eat aphids

Key points

● Intensive farming uses chemicals to produce crops as efficiently as possible. Organic farming does not use artificial chemicals to produce crops.

● The rate of photosynthesis is affected by light intensity, water availability, carbon dioxide concentration and temperature.

● Organic farmers add nutrients to soils using manure and compost. They kill pests using biological control, and remove weeds by hand or using a machine.

Summary questions

1 Copy and complete the following summary table to describe how farmers treat potential crop-growing problems:

Farming problem	Intensive farmers	Organic farmers
Pests		
Weeds		
Fungi		

2 Describe three methods organic farmers use to ensure their crops receive an adequate supply of nutrients.

3 Draw a table comparing the advantages and disadvantages of producing crops intensively and organically.

4.8

The use of chemicals in intensive farming

- What chemicals do intensive farmers use to increase crop yields?
- How are fertilisers manufactured?
- How are agricultural chemicals produced efficiently?

To produce large crop yields at low costs, intensive farmers rely heavily on the use of chemicals. Many agricultural scientists are employed to produce these chemicals. It is important that agricultural chemicals are produced efficiently – from both economic and environmental points of view. Examples of the types of chemical which intensive farmers use include:

- Pesticides – which kill pests. This reduces the risk of the crop being eaten.
- Fungicides – which kill fungi. This helps to protect the plant from disease.
- Herbicides – which kill weeds. This reduces competition from other plants.
- Fertilisers – these chemicals add essential nutrients to the soil.

   a List four types of chemicals used by intensive farmers.

Fertilisers

As a crop grows, it removes nutrients from the soil. To replace these nutrients, intensive farmers spray artificial fertilisers onto their fields. These contain soluble chemical compounds which the plants need to grow effectively. When it rains, these soluble compounds dissolve. They are then available for plants to absorb.

For healthy growth, plants need four important minerals – **nitrates**, **phosphates**, **potassium** and **magnesium**. The most widely used fertiliser in the UK is known as 'NPK fertiliser'. These fertilisers contain:

- nitrates – as a source of nitrogen (chemical symbol N)
- phosphates – as a source of phosphorus (chemical symbol P)
- potassium (chemical symbol K).

   b What are the three plant nutrients found in NPK fertiliser?

Figure 1 This farmer is spraying a ploughed field with chemical fertiliser

∞ links

For more information about the nutrients a plant needs to grow efficiently, see 4.11 Investigating plant growth.

Neutralisation

Chemical compounds can be defined as acidic, alkaline or neutral. The strength of an acid or alkali can be measured using the pH scale (see Figure 3). Universal indicator is often used to detect the pH of a chemical compound.

   c What is the typical pH of:

   i a strong acid   **ii** a weak alkali   **iii** a neutral solution?

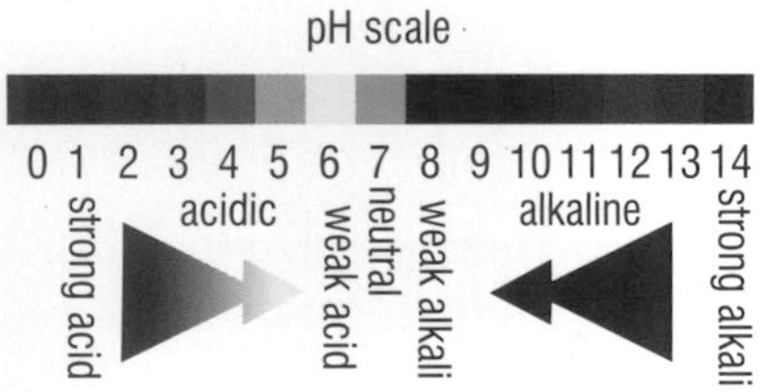

If an acid and an alkali are mixed together in the correct volumes and concentrations, a neutral product can be formed. This type of reaction is known as a neutralisation reaction.

Agricultural scientists use this idea to manufacture fertilisers. An example is ammonium nitrate, which is formed from reacting nitric acid with ammonia solution (an alkali):

Word equation: nitric acid + ammonia $\longrightarrow$ ammonium nitrate

Symbol equation: $HNO_3(aq)$ + $NH_3(aq)$ $\longrightarrow$ $NH_4NO_3(aq)$

Note that the compound formed is an aqueous solution – that is, it is dissolved in water.

Figure 2 NPK fertiliser

pH scale

0 1 2 3 4 5 6 7 8 9 10 11 12 13 14
acidic   alkaline
strong acid
weak acid
neutral
weak alkali
strong alkali

Figure 3 The pH scale, with universal indicator colours

Practical

Making ammonium sulfate fertiliser

Ammonium sulfate is a fertiliser which can be made in the laboratory. It uses a **neutralisation reaction** between ammonia solution and sulfuric acid:

ammonium hydroxide + sulfuric acid $\longrightarrow$ ammonium sulfate + water
(ammonia solution)

$$2NH_4OH(aq) + H_2SO_4(aq) \longrightarrow (NH_4)_2SO_4(aq) + 2H_2O(l)$$

Add 25 cm³ of ammonia solution to a conical flask. Using a burette, add dilute sulfuric acid until the ammonia solution becomes neutralised (check the pH of the solution using indicator paper). Allow the solution to crystallise – a residue of ammonium sulfate crystals will form over time.

The crystals formed can now be used as a fertiliser.

Safety: Wear eye protection

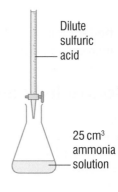

Dilute sulfuric acid

25 cm³ ammonia solution

Figure 4

Producing chemicals efficiently

It is important that agricultural chemicals are produced as efficiently as possible. This means that raw materials are not wasted, and the cost of these chemicals to farmers is as low as possible. This helps to ensure that the cost of intensively produced food is kept low. To make chemical processes as efficient as possible, agricultural scientists:

● build chemical plants near to raw materials
● recycle any chemical by-products
● ensure a good rate of chemical reaction
● maximise the yield from a reaction.

d State four ways chemical processes can be made as efficient as possible.

Summary questions

1 Match the chemicals used by intensive farmers to their use:

   FERTILISER To remove plants which could compete with the crop

   PESTICIDE To remove microorganisms which could cause crop disease

   HERBICIDE To add essential nutrients to soil

   FUNGICIDE To kill organisms which may eat a crop

2 **a** Describe what is meant by a 'neutralisation reaction'.
   b Write a word equation to describe a neutralisation reaction which could be used to manufacture a fertiliser.
   c Include a balanced chemical equation in your answer to part **b**. [H]

3 'It is important that chemicals are manufactured as efficiently as possible'. Discuss this statement from:
   a an economic viewpoint
   b an environmental viewpoint.

⃝⃝ links

For more information about the rate of chemical reactions, and maximising chemical yields, see 4.12 Rates of reaction, 4.13 Chemical yields and 4.14 Reversible reactions.

?? Did you know …?

Examples of chemical compounds which could be used to supply each of the plant nutrients N, P, K are:

Nitrates: Ammonium nitrate, NH_4NO_3

Phosphates: Calcium phosphate, $Ca_3(PO_4)_2$

Potassium: Potassium chloride, KCl

Key points

● Intensive farmers use pesticides, herbicides, fungicides and artificial fertilisers to maximise crop yields.

● Neutralisation reactions occur when acidic and alkaline solutions react together, to form a neutral product.

● Fertilisers can be made using the neutralisation reaction between ammonia solution, and either nitric acid or sulfuric acid.

● Agricultural chemicals are produced efficiently by: building chemical plants near the raw materials, recycling chemical by-products, ensuring a good rate of chemical reaction, maximising the yield from the reaction.

4.9 Rearing animals

Learning objectives

- How are animals reared intensively?
- What is meant by a controlled environment?
- How are animals reared organically?

AQA *Examiner's tip*

If you are asked to compare the two types of farming, make sure you stick to the facts. Answers such as 'Intensive farming is more cruel.' will not gain credit. To gain a mark you could say: 'Animals are kept in smaller spaces in intensive farming.'

Figure 1 Intensively farmed pigs are kept in highly controlled environments. In these batteries pigs are kept close together to keep warm, they are protected from predators, and fed a high-protein diet. This results in cheaper meat, but pigs are not able to display natural behaviours such as rolling in dirt.

Figure 2 Organically reared pigs have space to roam. They will not put on weight as quickly as intensively farmed pigs, therefore a farmer will need to feed and keep them for longer. This results in the meat being more expensive.

The combined work of agricultural and food scientists has resulted in:

- new, more intensive farming practices – higher inputs, such as chemicals, and additional heating and light, are used to produce higher animal or crop yields
- technological and scientific developments in growing crops and rearing animals – for example, resulting in higher yields and crops being able to be grown over a longer season
- better methods of storage, refrigeration and transportation of food – increasing a food's 'shelf-life'.

Controlled environments

Intensively farmed animals are kept in a strictly **controlled environment**, which makes the animals increase in size quickly. This makes intensively farmed animals and their products cheaper. However, some people raise concerns over the animals' wellbeing. They do not like the conditions the animals are kept in.

Many farmers would like to rear animals organically. However, animals farmed organically need more space, more time and more labour to look after them. This means costs are higher, which many farmers cannot afford.

The table below summarises the main differences between intensive and organic methods of animal production.

Factors that can be controlled	Intensively reared animals	Organically reared animals
Food supply	Animals are fed a high-protein diet to rapidly increase their body mass	Organic food is fed to the animals – for example, organically grown hay, grass and silage are often used as a food supply for cattle
Temperature	Animals are kept indoors, in a warm environment. Animals waste less energy heating their own bodies	Animals would normally live outdoors in the day time. At night or in bad weather, animals may be kept inside
Space	Restricted movement. Animals do not waste energy moving around	Animals are allowed to roam as freely as possible
Use of drugs	Antibiotics are given regularly to animals, to prevent the spread of disease	Antibiotics are not used, unless an animal is ill
Safety of enclosure	Animals are kept safe from predators	Increased risk from predators
Infection control	Animals are close-packed, so infections could spread quickly. Animals are checked regularly. Enclosures are under strict quarantine	Animals kept outside are more likely to catch infections from wild animals
Waste efficiency	Animal waste can be collected and converted into biogas	Animal waste is not available for biogas

a Why do intensive farmers give antibiotics to healthy animals?

b How is the amount of space an animal is provided with different in intensive and organic farming?

Battery-farmed versus free-range egg production

Around two-thirds of the total UK egg-laying hen population is kept in battery cages. On average, these hens will each lay 300 eggs a year. That is over 10 times the number produced by a wild hen! However, the hens spend their whole lives indoors, in cramped conditions.

Battery-farmed hens' living conditions

- Cages are stacked in large windowless sheds which are kept pleasantly warm. These sheds often accommodate more than 20 000 birds. A typical cage contains four or five birds. The minimum area allowed per bird is 550 cm^2. This is less than the size of a sheet of A4 paper.
- Food and water supplies are automated.
- Light conditions are often altered to depict spring, with long daylight hours. This is when hens naturally lay more eggs.
- Faeces drop through the bottom of the cages, preventing disease.
- Battery hens often suffer from foot deformities caused by the absence of suitable perches and restrictions of movement.
- Birds are prone to multiple fractures, caused by bone weakness. This is due to the high rate of egg production, resulting in calcium deficiency.

c Explain why hens kept in battery conditions can suffer from broken bones.

Organically reared hens' living conditions

- Must have at least an acre of field for every 400 chickens.
- Roam freely over pasture during daylight hours, but may choose to shelter indoors in cold, wet or windy weather.
- Inside space allows hens to move around and perch.
- Specially developed nest boxes give the birds the quiet and security they need to lay. They also ensure that the eggs can be collected quickly.

Figure 3 Battery-farmed hens are kept in a carefully controlled environment, which significantly increases egg production. Unfortunately it also reduces the hen's life expectancy.

Figure 4 These hens are roaming free on a farm. They will produce fewer eggs than battery-farmed hens.

Activity

Organic or Intensive farming

Write a balanced view (using arguments for and against) of these two types of farming.

Summary questions

1 Copy and complete the following table using the correct words:

Factor	Battery hens	Free-range hens
Space available	Large/Small	Large/Small
Access to outdoor space	Yes/No	Yes/No
Life expectancy	High/Low	High/Low
Number of eggs produced	More/Less	More/Less
Cost of eggs	High/Low	High/Low

2 Why do you think free-range eggs cost up to twice as much as battery-farmed eggs?

3 Explain the factors that an intensive farmer could control to increase the rate of meat production in pigs.

Key points

- Intensively farmed animals are kept in a strictly controlled environment. This makes the animals increase in size quickly.
- Factors that are controlled include their diet, size and temperature of enclosure and the routine use of drugs.
- Organically farmed animals are free to roam in large enclosures and are only given drugs when ill.

The impact of intensive farming on the environment

Learning objectives

- How does intensive farming affect the environment?
- How does intensive farming affect other animals?

∞ links

For information on the benefits of farming intensively look back at 4.7 Growing crops and 4.9 Rearing animals.

Figure 1 This algal bloom was photographed in marshland by the Thames Estuary. It is caused by fertiliser running off the land, into water.

Figure 2 This barren hillside has been severely affected by soil erosion, as a result of over-grazing by farm animals

Intensive farming produces as much food as possible, by making the best use of land, plants and animals. Intensive farmers maximise their outputs by various methods. These include breeding animals to maximise growth rates, using controlled environments, and using a number of different chemicals.

Intensive farming can have a harmful effect on the surrounding environment. This is caused by:

- **monoculture** – growing the same crop in an area of land for many years
- larger field sizes
- the use of chemicals – fertilisers, pesticides, herbicides and fungicides.

a What are the main four types of chemical used in intensive farming?

River pollution

When farmers apply fertiliser to their land, some may be wasted. This dissolves in rainwater and drains from the soil into ponds, lakes and rivers. The fertiliser in the water causes algae to grow rapidly, covering the water surface (see Figure 1). This stops light reaching the lower plants. These dying plants are broken down by bacteria in the water. The decaying process uses up lots of oxygen from the water, making it difficult for animals to survive. Many fish will die. This process is called **eutrophication**.

If high levels of nitrates (present in fertiliser) enter our drinking water, it can damage our health. Bacteria change the nitrates to nitrites. These can stick to red blood cells, stopping them from carrying oxygen.

b Why do fish and other animals die when fertilisers pollute waterways?

Soil erosion

Over-grazing of land leaves an area at risk of soil erosion (see Figure 2). This is a particular problem on hillsides, which are fully exposed to wind and to rain as it runs downhill. Over-grazing results in the loss of protective vegetation – mainly grass. Grass roots play an important role in binding soil together. Barren soil also dries out easily. This means it can be easily eroded by wind or water.

Many intensive farmers have removed hedgerows to create larger fields. This creates more space for growing crops. The land also becomes easier to farm as large machinery can be used. However, hedges are homes for hundreds of plant and animal species. Removing them destroys many organisms' habitats. It also causes soil erosion, as the hedges act as natural wind breaks. Once hedges are removed, soil can be blown or washed away.

c Which natural processes erode soil?

Poisoning wildlife

DDT is a very effective pesticide – only very small amounts are needed to kill an insect. DDT was a widely used pesticide, which saved many people from starvation by killing crop pests. It has also been used to kill the mosquitoes that spread malaria.

However, DDT has killed large numbers of wildlife, as it is toxic at high concentrations. DDT does not decompose easily. Because of this, it passes along a food chain until it reaches fatal levels – killing many top predators (see Figure 3). As a result the UK (and many other countries) have banned the use of DDT.

d State an advantage and a disadvantage of spraying crops with DDT.

Problems with monoculture

Monoculture means growing one crop on an area of land. This results in the effects below.

- The same nutrients are constantly removed from the soil. Farmers overcome this problem by using large amounts of fertiliser.
- Crop pests are likely to be concentrated in one area. This is a particular problem if land is to be used for the same crop year after year. It leads to a greater use of pesticides.

By removing the mix of plants that once grew in an area, the habitat for some insects and birds that used to feed on these plants has gone. This reduces **biodiversity**.

e What does monoculture mean?

Food production and distribution

Food is often consumed far from where it is produced. For example, much of our rice comes from Thailand, and many bananas come from the Caribbean. This adds to the cost of the product. It also has a negative effect on the environment. Fuels are burnt while transporting the product, adding to the world's carbon dioxide emissions. 'Food miles' is the term often used to describe the distance travelled by a food product from producer to consumer.

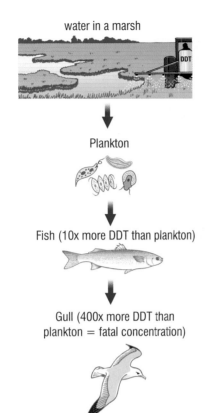

Figure 3 DDT accumulation in a food chain

water in a marsh

Plankton

Fish (10x more DDT than plankton)

Gull (400x more DDT than plankton = fatal concentration)

Activity

Ecological footprint – how big is yours?

An 'ecological footprint' measures how much land and water area a human being requires, to produce the resources they consume (food and energy, for example), and to absorb the waste they produce.

- Try calculating your ecological footprint to see if your current lifestyle is sustainable. What simple steps could be taken to reduce your impact on the environment?

Summary questions

1 Match the intensive-farming technique to its potential environmental effect:

Adding fertiliser	Untargeted animals being killed
Removal of hedges	River pollution and fish dying
Use of pesticides	Soil erosion

2 State two environmental consequences of removing hedges to create larger fields.

3 Explain why fish can die when a river gets polluted with fertiliser.

Key points

- Intensive farming can pollute rivers and lakes, cause soil erosion, and result in the removal of nutrients from soil.

- Chemicals used in intensive farming can kill untargeted animals.

- Animals may also lose their habitats when hedgerows are removed.

4.11 Investigating plant growth

Learning objectives

● How can you measure plant growth?

● How are plants grown hydroponically?

● What factors affect the rate of plant growth?

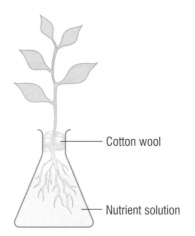

Figure 1 A simple example of a hydroponic culture

Figure 2 These researchers are using a ruler and a laptop computer in a crop size experiment. They are investigating the effects of herbicides and fertilisers.

Hydroponics

To ensure that a plant receives the minerals it needs, it can be grown in water with minerals dissolved in it. This system of growing plants is known as **hydroponics**. This is a technique often used by commercial growers, because it enables plants to grow quickly.

Growing plants under hydroponic conditions is a useful tool to enable agricultural scientists to study plant growth. By growing a plant hydroponically, all factors which may affect plant growth can be controlled.

Practical

Standard procedure – What affects the rate of plant growth?

Plant growth is affected by a number of factors. These include the nutrient conditions, light conditions, amount of water supplied, or overcrowding. In this investigation, you will select *one* factor to investigate.

1 Choose a factor to investigate.

2 Set up six plants in hydroponic culture. One will act as a control, the other five will have varied conditions for the factor being investigated.

● Germinate several seeds in damp, cotton wool.

● When the seedlings are about 5 cm tall, transfer them to conical flasks containing a nutrient solution. (See Figure 1.)

● Add a wick between the solution and the cotton wool to keep the cotton wool moist, until the roots of the plant reach the nutrient solution.

3 Place seedlings in the conditions selected for your investigation.

Ensure all factors are kept constant, except for the one you are studying. You should allow one seedling to have optimum levels of all four factors to act as a control. These are possible ways you could vary each factor:

● Light intensity

Place specimens different distances from a light source. For quantitative data, a light intensity meter should be used at each location. This can measure the intensity of the light reaching the plant.

● Overcrowding

Add different numbers of specimens to each container.

● Nutrient availability

Use flasks containing different nutrient solutions. The control should contain all four nutrients (nitrate, phosphate, potassium and magnesium).

● Amount of water supplied

This factor should **not** be investigated hydroponically, as this growing technique involves an excess of water! Plant in compost, and add different volumes of water to each seedling.

4 Measure plant growth at fixed intervals

Decide how you will assess plant growth. You might choose to measure plant height, root length, plant mass, number of leaves or average leaf size.

Factors affecting plant growth

Agricultural scientists study factors which affect plant growth, to develop optimum growing conditions. This increases a plant's growing season and speed of growth, resulting in higher yields. These include:

Light availability – Plants require an intense source of light, to ensure that they photosynthesise rapidly. The glucose produced is used for growth, resulting in high crop yields. Some farmers use artificial lights so that photosynthesis can occur for longer. This can be used to produce 'summer conditions' all year around.

Water availability – Most growing plants contain as much as 90% water. For optimum growth, plants need an adequate water supply. This allows the plant to photosynthesise rapidly. It also ensures the structure of the plant is supported. If a plant lacks water, it will wilt as there is not enough water in its cells.

Space – Plants need enough space to absorb sunlight, and nutrients and water from the soil. If plants become overcrowded, they tend to be smaller and produce lower yields.

Nutrient availability – Plants need minerals for healthy growth. These are normally acquired from the soil.

a How can you tell if a plant is lacking in water?

Role of nutrients in growth

Plants need minerals for healthy growth. These are normally taken in from the soil. When crops are harvested, minerals are removed from the ground. Farmers need to be able to recognise mineral deficiency symptoms in plants so that they can add appropriate chemicals – normally fertilisers.

Mineral	Role in plant		Appearance of a deficient plant
Nitrates (contain **nitrogen**)	Nitrates are involved in making DNA and amino acids. The amino acids join together to form proteins that are needed for healthy leaf growth.		Older leaves are yellowed, growth is stunted
Phosphates (contain **phosphorus**)	Needed for healthy roots		Younger leaves have a purple tinge and poor root growth
Potassium	Needed for healthy leaves, flowers and a high fruit yield		Yellow leaves with dead areas on them
Magnesium	Needed for making chlorophyll molecules, essential for photosynthesis		Leaves turn pale and then yellow

b How could you tell a plant was lacking in magnesium?

c How could farmers replace nutrients that are lacking in their soil?

Summary questions

1 Name three factors you can measure to monitor plant growth.

2 How are plants grown hydroponically?

3 Make a list of the factors that can affect the growth of a plant.

Key points

- Plant growth can be measured in a number of ways, including height and mass of plant or size and number of leaves.
- Plants grown hydroponically are grown in a solution of water containing dissolved minerals.
- Some of the main factors affecting plant growth are light, water, space and availability of nutrients.

4.12 | Rates of reaction

Physical and chemical changes

Physical changes are temporary changes, which are usually easy to reverse. No new substances are made. One example is when salt dissolves in water.

Chemical changes are more permanent changes, which are often very difficult to reverse. For example, when acid rain reacts with marble, releasing carbon dioxide gas. Chemical changes (chemical reactions) involve the breaking and making of chemical bonds, producing new substances called **products**.

An example of a useful product formed from a chemical reaction is artificial fertiliser. One type of fertiliser is ammonium nitrate. This is formed from the reaction between ammonia and nitric acid.

   a What is the difference between a physical change and a chemical reaction?

   b Give the name of a chemical which can act as a fertiliser.

Collision theory

The substances that react together are called **reactants**. Their particles may be atoms, ions or molecules. In order for a chemical reaction to occur, the particles must (i) collide, and (ii) collide with enough energy to cause a reaction. The more often particles of the reactants collide together, and the greater the energy of those collisions, the faster the chemical reaction will happen.

Figure 1 Chemical reactions occur when particles of the reactants collide together

Affecting the rate of a chemical reaction

Agricultural scientists produce a number of chemical products. These include pesticides, herbicides, fungicides and fertilisers. By understanding how chemical reactions can be controlled, agricultural chemicals can be produced as efficiently as possible. This means that food products are produced more cheaply, benefiting the consumer.

Several factors affect the rate of a chemical reaction. These include:

Concentration

The more concentrated a solution becomes, the more dissolved particles there are in the same volume of solution. This means that it is more likely that a collision will occur between reactant particles. The greater the chance of a collision, the greater the rate of reaction.

An increase in concentration increases the number of collisions in a given time, and so increases the rate of reaction.

Less concentrated solution   More concentrated solution

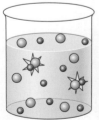 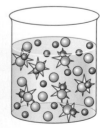

Figure 2 More concentrated solutions have more dissolved particles in the same volume of liquid. So particles collide more often.

Temperature

When the temperature of a substance is increased, the energy of its particles increases. This means that the particles move about more quickly. This affects the rate of reaction in two ways:

- More frequent collisions between reactant particles occur. This increases the rate of reaction.

- Sometimes when a collision occurs, the particles do not have enough energy to react. The particles just 'bounce off' each other. When temperature is increased more particles have enough energy to react, so more collisions result in a reaction. This increases the rate of reaction.

The minimum energy required for a reaction to occur is known as the **activation energy**.

Surface area

If a solid reacts with a liquid, collisions between reactants can only occur on the outer surface of the solid. If the solid is broken into smaller pieces, the surface area increases. More particles on the surface of the solid are available to collide with other reactant particles. This means more collisions can occur in a given time between reactant particles.

An increase in surface area increases the rate of reaction.

Use of catalysts

Catalysts are substances which are added to a chemical reaction. Their presence speeds up the chemical reaction, but the catalyst does not get used up in the reaction. Catalysts work by reducing the activation energy required in a chemical reaction.

Adding a catalyst increases the rate of reaction.

c State the four factors which can affect the rate of a chemical reaction.

Increasing the rate of a chemical reaction

Industrial chemists use their knowledge of factors which affect chemical reactions to ensure that chemicals are manufactured as quickly and efficiently as possible. Typical conditions to increase the rate of reaction at a chemical manufacturing site would include:

- the use of concentrated solutions
- solid reactants made into small pieces
- high temperatures in the reaction vessel
- use of a catalyst for reactions whenever possible.

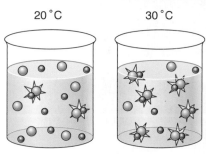

Figure 3 At higher temperatures, particles move more quickly, so particles collide more often. The chance of a collision having enough energy to cause a reaction also increases.

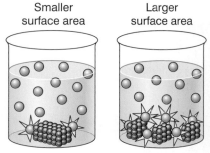

Figure 4 When a solid is broken into smaller pieces, its surface area increases. This causes the rate of reaction to increase.

??? Did you know …?

Enzymes are biological catalysts. They speed up many chemical reactions which occur in the body including digestion.

Key points

- When particles of reactants collide together with sufficient energy, they react to produce a new product. This is known as 'collision theory'.

- The rate of a chemical reaction can be affected by: concentration, temperature, surface area and the use of a catalyst.

- Chemical reactions can be speeded up by: increasing the concentration of a solution; increasing the temperature; increasing the surface area of a solid; using a catalyst.

Summary questions

1 Link the factor to its effect:

Factor	Effect
Concentration	Increases the energy of the particles
Temperature	Decreases the activation energy of a reaction
Surface area	Increases the number of dissolved particles in a given volume of solution
Catalyst	Increases the number of solid particles exposed

2 Explain why increasing the temperature of the reactants affects the rate of reaction in two different ways.

3 'It is not always economically viable to increase the temperature of a chemical reaction.' 'Increasing the temperature increases the rate of a chemical reaction.' Explain how both of these statements can be true.

4.13   Chemical yields

Learning objectives

- What is meant by a yield?
- What are theoretical and actual yields?
- How can you calculate percentage yield?

Figure 1 Calcium oxide is used by farmers to neutralise (or raise the pH of) acidic soils. It is produced on a large scale in many countries.

⊂⊃ links

For information on using balanced equations for chemical reactions see 5.6 Balanced equations.

When two reactants combine together to form a product, the amount of useful product made is known as the **yield**. For example, calcium oxide is used by farmers to neutralise (or raise the pH of) acidic soils. What yield could be produced from 100 tonnes of calcium carbonate?

Word equation: calcium carbonate $\longrightarrow$ calcium oxide + carbon dioxide
                               (limestone)

Symbol equation: $CaCO_3(s)$ $\longrightarrow$ $CaO(s)$ + $CO_2(g)$

Masses (tonnes):   100 $\longrightarrow$ 56   +   44

(See Maths skills box below for how to work out the numbers.)

The yield from 100 tonnes of calcium carbonate in this chemical reaction is 56 tonnes of calcium oxide.

Theoretical yield

The yield which a chemical reaction should produce is known as the **theoretical yield**. If there is only one product, this can be calculated by adding the masses of the reactants involved. Remember that mass is always conserved in a chemical reaction:

total mass of products = total mass of reactants

You can check this in the reaction above.

Maths skills

Calculating a theoretical yield

The theoretical yield can be calculated using five steps as follows:

Worked example

Calcium oxide is an important agricultural chemical. It is produced through the thermal decomposition of calcium carbonate (in limestone). What is the maximum mass of calcium oxide that could be produced from 500 tonnes of calcium carbonate?

Step 1 Write down the balanced chemical equation:

Word equation: calcium carbonate $\longrightarrow$ calcium oxide + carbon dioxide

Symbol equation: $CaCO_3(s)$ $\longrightarrow$ $CaO(s)$ + $CO_2(g)$

Step 2 Work out the relative formula masses of the reactants and products:

Molecule	Formula	Atoms present	Atomic mass × Number of atoms	Relative formula mass
Calcium carbonate	$CaCO_3$	1 calcium atom 1 carbon atom 3 oxygen atoms	40×1 12×1 16×3	100
Calcium oxide	CaO	1 calcium atom 1 oxygen atom	40×1 16×1	56
Carbon dioxide	CO_2	1 carbon atom 2 oxygen atoms	12×1 16×2	44

Step 3 Work out the masses of the reactants and products from the equation:

Equation	$CaCO_3$ (s) $\longrightarrow$ CaO (s)	+	CO_2 (g)
Mass from equation	100	56	44

Step 4 Work out the multiplication factor for this reaction:

Multiplication factor $= \dfrac{\text{actual mass of reactant}}{\text{relative formula mass of reactant}} = \dfrac{500}{100} = 5$

Step 5 Calculate the theoretical yield:

Apply the multiplication factor to Step 3.

Equation	$CaCO_3$ (s)	+	CaO (s) $\rightleftharpoons$	CO_2 (g)
Actual mass (kg)	$5 \times 100 =$ 500 tonnes		$5 \times 56 =$ 280 tonnes	$5 \times 44 =$ 220 tonnes

The **theoretical yield** of calcium oxide which can be produced from 500 tonnes of calcium carbonate is **280 tonnes**.

AQA Examiner's tip

Follow the five-step method to calculate theoretical yield:
1 Write down the balanced chemical equation.
2 Work out the relative formula masses of the reactants and products.
3 Work out the masses of the reactants and products from the equation.
4 Work out the multiplication factor for this reaction.
5 Calculate the theoretical yield.

Actual yield

In reality, theoretical yields are never reached. The amount of product you actually get out of a reaction is known as the **actual yield**. The actual yield is lower than the theoretical yield because:

● Some of the reactants may not react or may not be pure.
● Some of the product may be left behind in the reaction vessel or lost in processing.
● Some chemical reactions are reversible.

Percentage yield

Agricultural scientists can evaluate the most efficient way of producing a chemical, by comparing the **percentage yield** of different manufacturing processes.

For example, calcium oxide is manufactured in a lime kiln. When 500 tonnes of calcium carbonate are added, the theoretical yield is 280 tonnes. In practice, only 252 tonnes of calcium oxide can actually be collected. The percentage yield of this reaction is –

Percentage yield $= \dfrac{\text{actual yield}}{\text{theoretical yield}} \times 100 = \dfrac{252}{280} \times 100 = \textbf{90\%}$

Summary questions

1 Define the terms: **a** theoretical yield, **b** actual yield, **c** percentage yield.

2 Calcium oxide is an important agricultural chemical, used to increase the pH of soil. It is produced by the thermal decomposition of calcium carbonate (in limestone), according to the reaction:

$CaCO_3 \longrightarrow CaO + CO_2$

What is the theoretical yield of calcium oxide which could be produced from 400 kg of calcium carbonate?

3 In the manufacture of ammonium nitrate fertiliser, an actual yield of 160 kg is produced when 68 kg of ammonia (NH_3) are added to excess acid (HNO_3). What is the percentage yield for this reaction?

$NH_3 + HNO_3 \longrightarrow NH_4NO_3$

Key points

● The amount of product collected from a chemical reaction is known as the yield.

● The theoretical yield is the maximum amount of product a reaction could produce; the actual yield is the amount of product which is collected from a reaction.

● Percentage yield is the proportion of a product collected compared to the theoretical yield, expressed as a percentage.

4.14

Reversible reactions

Learning objectives

- What is a reversible reaction? [H]
- How is ammonia manufactured? [H]
- How do temperature and pressure affect yield? [H]

Some chemical reactions release energy – these are known as **exothermic** reactions. Others take in energy during the reaction – these are **endothermic** reactions.

a What is the difference between an exothermic and an endothermic reaction?

Many chemical reactions are 'one-way' reactions. Once the product is formed, the reaction is complete. However, some reactions are 'two-way': these are known as reversible reactions. In a reversible reaction, the products can react together to re-form the reactants.

Reversible reactions can be represented in the following way:

$$A + B \rightleftharpoons C + D$$
$$\text{reactants} \rightleftharpoons \text{products}$$

One reaction direction is exothermic, the other endothermic.

b What is meant by a 'reversible reaction'?

The manufacture of ammonia

Ammonia forms when gaseous nitrogen and hydrogen react together. This is an example of a reversible reaction. Some of the ammonia produced dissociates (splits) back into nitrogen and hydrogen:

exothermic

Word equation: nitrogen + hydrogen $\rightleftharpoons$ ammonia

Symbol equation: $N_2(g)$ + $3H_2(g)$ $\rightleftharpoons$ $2NH_3(g)$

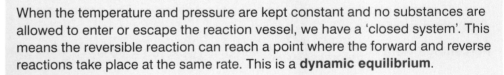
endothermic

When the temperature and pressure are kept constant and no substances are allowed to enter or escape the reaction vessel, we have a 'closed system'. This means the reversible reaction can reach a point where the forward and reverse reactions take place at the same rate. This is a **dynamic equilibrium**.

c What is meant by the term 'dynamic equilibrium'?

Manufacturing ammonia efficiently is difficult! At low temperatures, few particles have the required energy needed to react. The reaction takes place extremely slowly. Increasing the temperature causes the reaction to take place at a faster rate. However, increasing the temperature also supplies energy to the reverse (endothermic) reaction, and so more ammonia dissociates back into the reactants. The equilibrium point is shifted left towards the reactants – the yield decreases. But the need to produce ammonia quickly in industry means a compromise temperature of 450 °C is chosen.

Increasing the pressure in the reaction vessel also affects the rate and yield of the reversible reaction. This causes the equilibrium point to shift right, towards the product. The yield increases, as does the rate of reaction.

Figure 1 Ammonia manufacturing plant. Ammonia is an important agricultural chemical. It is used in the manufacture of fertilisers; over 80% of the world's production of ammonia is used in this way.

Conditions affecting the yield of ammonia:

For all reversible reactions, the yield is affected in the following ways:

Factor	Effect
Increasing temperature	Equilibrium shifts in favour of the endothermic reaction direction
Increasing pressure	Equilibrium shifts in favour of the side with fewer molecules of gas

$N_2(g) + 3H_2(g) \rightleftharpoons 2NH_3(g)$ High temperature – equilibrium favours reactants

$N_2(g) + 3H_2(g) \rightleftharpoons 2NH_3(g)$ High pressure – equilibrium favours products

Increasing the temperature shifts the equilibrium in favour of the reactants. The yield decreases. Increasing the pressure shifts the equilibrium in favour of the product. The yield increases.

Did you know ...?

Under the conditions chosen for the Haber process, ammonia can only be produced with a percentage yield of around 15%.

The Haber process

Ammonia is produced using the Haber process. The optimum conditions required to manufacture ammonia have been carefully researched. A balance needs to be found between producing a high yield of ammonia, and the costs of producing these reaction conditions. Here are the conditions chosen:

- A temperature of around 450 °C.
- A pressure of around 200 atmospheres.
- The use of an iron catalyst.

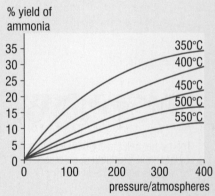

Figure 2 Graph showing how altering temperature and pressure affects the yield of ammonia

d List the typical conditions used in the Haber process.

e Using the graph in Figure 2, what would be the percentage yield of ammonia produced at a temperature 500 °C, and a pressure of 200 atmospheres?

Key points

- Reversible reactions are chemical reactions which can take place in two directions. [H]

- Ammonia is manufactured using the Haber process. The following conditions are required in this process: 450°C, 200 atmospheres pressure and an iron catalyst. [H]

- For all reversible reactions: increasing temperature favours the endothermic reaction; increasing pressure favours the side of the reaction with fewer molecules of gas. [H]

Summary questions

1 Copy and complete the sentences below, using the following words:

endothermic temperature reversible decrease increases pressure

Some reactions are .... – these are reactions that can operate in two directions. One reaction direction is an exothermic reaction – the other is .... In the Haber process, increasing the .... favours the endothermic reaction direction. This causes the yield of ammonia to .... Increasing the .... causes the equilibrium to shift in favour of the product. This .... the yield of ammonia. [H]

2 Why is a pressure of greater than 300 atmospheres not used in the Haber process? [H]

3 Explain why, during the manufacture of ammonia, increasing the temperature increases the rate of reaction, but decreases the yield of the product. [H]

<table>
<tr><td>

4.15

</td><td>

Selective breeding and genetic engineering (k)

</td></tr>
</table>

Learning objectives

- What is selective breeding?
- What is genetic engineering? [H]
- What are the advantages of producing organisms through genetic engineering rather than selective breeding? [H]

Farmers select the animals they rear, or plants they grow, by their characteristics. These characteristics are beneficial to the farmer. For example, they choose sheep that produce lots of wool, or chickens that lay lots of eggs.

To ensure that they maintain their desired stock of plants or animals, farmers choose which plants or animals should mate. This is called **selective breeding**.

a What characteristics would a pig farmer select for?

Selectively breeding cattle

The type of cow a farmer desires depends on the produce they sell. Beef farmers need cows that can quickly turn grass and feed into meat, whereas dairy farmers need cows that have high milk yields (see Figure 1).

b How does a dairy farmer choose which offspring to breed?

Advantages and disadvantages of selective breeding

The farmer selects the cow with the highest milk production, he breeds this with his best bull.

OFFSPRING
(1st GENERATION)

The farmer chooses the best cow and breeds again with his best bull.

This process continues over several generations.

OFFSPRING (MANY GENERATIONS LATER)

All cows have the desired characteristic of high milk production.

Figure 1 Selectively bred dairy cattle

Advantages	Disadvantages
Higher yields	Plants without seeds can no longer reproduce naturally
Produce available for longer periods of the year	Reduces variation
Extends an organism's tolerance range – for example, increased resistance to certain diseases or ability to survive frosts	Some pedigree animals have poorer health
Produces a more uniform crop – this makes crops easier to harvest. For example, they grow at the same height, and are ready at the same time	Reduces the **gene pool** – if a new disease arises, an organism may not exist that contains the gene for resistance to this disease
Chooses the characteristics of the food item required – therefore producing an item more suited to people's requirements	

Genetic engineering

Producing animals with desired characteristics through selective breeding is a slow process. It is also not very accurate. Farmers were not originally aware that in this process they were changing the organism's genes. Scientists now have a much greater understanding of genetics. They are able to alter an organism's genes to produce the desired characteristics. This is called **genetic engineering** (or **genetic modification**). It can happen in one generation.

c Name two advantages of genetic engineering over selective breeding.

Genes from another organism (foreign genes) are put into plant or animal cells at a very early stage in their development. As the organism develops, it will display the characteristics of the foreign genes.

Higher

Higher

d Why are the genes inserted into the plant or animal cells known as 'foreign genes'?

A useful gene is removed from the nucleus of a donor cell.

⬇

The foreign gene is then put into a circular piece of DNA called a plasmid. This is now known as a piece of recombinant DNA.

⬇

The recombinant DNA is put into a bacterial cell.

⬇

The bacteria reproduce lots of times, producing lots of copies of the recombinant DNA.

⬇

Plant cells are infected with the bacteria. The foreign gene becomes integrated with the DNA of the plant cells.

⬇

The plants cells are placed in a growing medium to grow into plants. These plants will have the desired characteristics.

Figure 2 Process of how tomatoes and other crops are genetically modified

Frost-resistant tomatoes are an example of a genetically modified crop. Scientists isolated a gene in flounder (a type of fish) which codes for an antifreeze chemical. This enables the fish to survive in very cold water. This gene was then inserted into a tomato plant. Animals are genetically modified in a similar way, by inserting the required genes into an embryo.

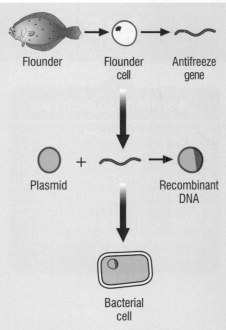

Flounder → Flounder cell → Antifreeze gene

Plasmid + 〜 → Recombinant DNA

Bacterial cell

Figure 3 First steps in producing genetically engineered frost-resistant tomatoes

??? Did you know ... ?

Three examples of genetically engineered organisms include:

● Cotton – GM cotton has a high yield and pest resistance

● *E. coli* – bacteria engineered to produce insulin

● Glo-Fish – a fish, engineered to glow when it comes into contact with environmental pollution.

Activity

Increasing yield

Carry out research into how farmers have used selective breeding and genetic engineering to increase the yield of plant and animal products. Discuss:

● How has selective breeding changed farming over time?

● How has genetic engineering changed farming over time? **[H]**

Summary questions

1 Copy and complete the following sentences, using the words below:

genetic engineering   selective   desired   offspring
organisms   foreign genes   characteristics

To produce an organism with .... characteristics, farmers choose .... with the best features. They then choose the best .... and breed them again. This is called .... breeding. Scientists can insert .... into an organism to change .... This is called ....

2 Name an example of a selectively bred:
  a animal   **b** plant.

3 Why must foreign genes be inserted in an embryo, and not at a later stage in an organism's development? **[H]**

Key points

● Farmers selectively breed plants and animals by breeding organisms with desired characteristics.

● Scientists are able to alter an organism's genes to produce organisms with desired characteristics – genetic engineering. **[H]**

● Genetic engineering is a faster and more accurate technique than selective breeding. **[H]**

Standard procedures used in food science

Food science is an applied science that studies the production, marketing and consumption of food.

Food scientists perform a wide variety of jobs. They include dieticians, nutritionists, microbiologists, public health inspectors, agricultural scientists and food analysts.

a Give three examples of jobs carried out by food scientists.

When do food scientists follow standard procedures?

Food scientists use a number of standard procedures to perform their role, including:

- analysing the quality of food products – to ensure that a product contains what the manufacturer claims
- checking the safety of food products – to check for the presence of microorganisms, and determine the type and number present
- applying scientific methods to keep food fresh, safe and attractive
- investigating new manufacturing methods to produce food more quickly and cheaply
- creating new food products
- ensuring that food labelling complies with legislation – for example, noting the presence of ingredients which may cause an allergic reaction.

b What is a standard procedure?

Activity

The perfect cup of tea

How can you be certain that someone will make a cup of tea in the way you like it? Often you may be offered a cup of tea that is too strong, too milky or too sweet. Write a standard procedure for producing your perfect cup of tea. Remember attention to detail is essential. For example, how much milk should be added, and how long should you let the tea bag remain in the water?

Qualitative and quantitative tests

Food scientists carry out a number of tests to determine what is in the foods and drinks we buy. Some of these detect the presence of a substance – this is called a **qualitative test**. It tells you whether or not the food or drink contains a specific substance such as an additive or an unwanted microorganism. To find out how much of the substance is present, a **quantitative test** must be used. For example, a **serial dilution** could be performed to estimate how many bacteria are present.

c Which type of test (qualitative or quantitative) would you perform if you wanted to find out how much sugar was in a chocolate bar?

Remembering standard procedures

Use the following table to help you to revise the standard procedures.

Students need to know:	Topic	Further guidance
How to use aseptic techniques to swab areas to detect the presence of bacteria	4.4	Wipe the swab across the surface being sampled, then lightly across a sterile agar plate. Incubate for 48 hours, then identify any organisms present.
How to complete serial dilutions to make accurate bacteria counts	4.4	Working aseptically, dilute the bacterial sample using sterile nutrient broth. Then place 1 cm³ of the final solution on an agar plate. Keep going until individual colonies can be seen and counted. Work out the number of bacteria present in the original sample using the formula: Number of bacteria (per cm³) in original sample = number of colonies × dilution of sample.
How to make streak plates to identify the type of bacteria present	4.4	Working aseptically, use a sterilised wire loop to make streaks on an agar plate. Flame the loop and allow to cool. Spread the bacteria about by streaking the plate in another direction. Repeat twice more. The species of bacteria can be identified by the appearance of its colonies.
How to carry out tests on food products to determine the level of bacteria in the food	4.4	Working aseptically, carry out a serial dilution as described above. Count how many of each type of colony are present. Then work out how many of each bacterial species were in the original sample using the formula: Number of bacteria (per cm³) in original sample = number of colonies × dilution of sample.
How to plan and assess how well a plant grows under various conditions	4.11	Plants should be grown hydroponically to ensure they all have access to the same nutrients and amount of water. One plant should be placed in each container to ensure space, and placed in a window sill to access light. You should then change one variable to study its effect on growth. All other factors must be kept constant.

Key points

- A standard procedure is a step by step method for carrying out a test. It ensures that anybody who performs the test will achieve the same results.

- Qualitative tests test for the presence of a substance. Quantitative tests determine how much of a substance is present.

- Food scientists monitor the contents, quality and safety of food products, as well as developing new products.

Summary questions

1 What is the difference between a qualitative and a quantitative test?

2 Why is it important to follow standard procedures?

3 Why is it important to carry out microbiological techniques under aseptic conditions?

Summary questions

1 a Microorganisms play an important role in the production of a number of food and drink products. Copy and complete the table, stating for each of the following products, whether they are produced using yeast or bacteria:

Product	Microorganism
Bread	
Cheese	
Yoghurt	
Wine	

b To make beer, brewers add yeast to the malt. What process does the yeast carry out to turn this into alcohol?

c Which gas is also made by this process?

d Describe how bread is made from flour, water and sugar.

2

a Which mineral is this plant deficient in?

b How do organic farmers increase the nutrient content of soil?

c Describe the process of eutrophication.

3 a What causes food poisoning?

b Describe the common symptoms of mild food poisoning.

c Name **two** practices you should follow at home, to prevent food from being contaminated with microorganisms.

d To prevent bacteria growing in food products, manufacturers either add substances to the food, or treat the food. Name two of these techniques and explain how they work.

4 a Rearrange the following sentences to explain how farmers may selectively breed their livestock.
1 Offspring grow.
2 Farmer chooses animals with the best characteristics.
3 Farmer then chooses offspring with the best characteristics.
4 These animals are then bred.
5 These are then bred. This process is repeated for many years.

b Give two disadvantages of selective breeding.

5 An agricultural scientist wishes to manufacture ammonium nitrate fertiliser. A new technique of producing this fertiliser is analysed. 34 kg of ammonia are reacted with excess nitric acid, according to the following reaction:

$$NH_3 + HNO_3 \longrightarrow NH_4NO_3$$

It is noted that 120 kg of fertiliser are produced by this reaction:

a What is the actual yield from this reaction?

b Calculate the theoretical yield from this reaction.

c Calculate the percentage yield from the reaction.

d Suggest why the actual yield is lower than the theoretical yield.

6 Discuss the reasons why the following conditions are used during the manufacture of ammonia: **[H]**

a temperature of around 400 °C

b pressure of around 200 atmospheres

c presence of an iron catalyst

7 Bt corn is an example of a genetically modified crop. Scientists isolated a gene from *Bacillus thuringiensis* (a type of soil bacteria) and put it into corn. This gene codes for a toxin which kills insects. This means that insecticides are not needed for this crop. Describe step by step how scientists genetically engineered corn to be insect resistant. **[H]**

Key practical questions

1 Describe how to:

a sterilise a wire loop

b detect the presence of microorganisms using a sterile swab

c make a streak plate.

2 Why do we :

a use aseptic technique when working with microorganisms

b use serial dilutions to count the number of bacteria in a sample

c grow plants hydroponically to assess how well a plant grows in the presence of different minerals?

AQA Examination-style questions

1 a Microorganisms can be used in the production of food.
   i Describe how microorganisms are used to produce yoghurt   **(2)**
   ii Why do bakers use yeast when making bread?   **(2)**
b This label was found on a packet of dried soup.

> To serve, empty sachet into a cup, add 250 ml of boiling water and stir well
>
> Ingredients:
> Water, Vegetables, Sugar, Starch, Vegetable oil, Salt, Monosodium glutamate, Wheat flour
> Store in a cool, dry place

Explain why this product should be kept in a cool dry place.   **(2)**

2 With intensive farming, different chemicals are added to the crops to improve crop yield.
a Describe how weeds prevent crops growing well.   **(1)**
b Name the type of chemical used to kill weeds in intensive farming.   **(1)**
c A farmer investigates the effectiveness of different herbicides. He records his yield when he uses the different herbicides in different fields.

	Herbicide			
	A	B	C	none
Yield (kg/acre)	2700	3200	2700	1500
Cost of herbicide (£/acre)	4.20	7.30	3.90	0

   i Explain why the farmer grew one field with no herbicide in this investigation.   **(1)**
   ii State which fertiliser the farmer should use and explain your answer.   **(3)**
d i State how an organic farmer would solve the problems caused by weeds.   **(1)**
   ii Organic farmers cannot use chemical pesticides. Give **one** way an organic farmer can prevent pests from destroying their crops.   **(1)**

3 The Food Standards Agency (FSA) is responsible for making sure that food is safe to eat and will not cause food poisoning.
a i Give an example of a bacterium that could cause food poisoning.   **(1)**
   ii The staff in a restaurant should be aware of the need for good personal hygiene. Give one precaution that they should take, and explain why.   **(2)**
   iii Describe how a Public Health Inspector would test samples taken from the kitchen surfaces for the presence of bacteria.   **(2)**

b The FSA often has to determine the number of bacteria in food. Milk is one example of a food that would be tested. Describe how to complete serial dilutions of milk and then do an accurate bacterial count.
You may use a diagram to help you with your answer.   **(4)**
c A food scientist at the FSA did some serial dilutions of a sample of milk containing bacteria. The final dilution was 1:1000.
The food scientist took 0.1 ml of the final dilution and grew the bacteria on an agar plate.

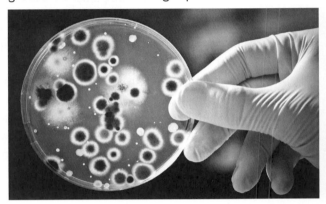

   i 30 Colonies of bacteria grew on the plate. Use the formula to work out the number of bacteria in the sample of milk.   **(1)**
Total bacterial count in 0.1 ml = number of colonies × dilution.
   ii How many bacteria would there be in 1 ml of the sample?   **(1)**
AQA, 2008

4 Many farmers keep chickens. Chickens can be reared on organic farms.

The table shows the numbers of chickens (in millions) reared on organic farms in the UK from 2004 – 2007.

Year	Number of organic chickens (in millions)
2004	2.2
2005	2.5
2006	3.5
2007	4.4

a i Use the data in the table to plot a bar chart.   **(3)**
   ii Describe what the bar chart shows.   **(1)**
   iii Suggest a reason for your answer to 4 a) ii).   **(1)**
b i Describe how chickens are reared on organic farms.   **(2)**
   ii Give **two** disadvantages to the **farmer** of rearing chickens on organic farms.   **(2)**
AQA, 2010

1 An agricultural scientist is investigating the growth of potato plants.

    a Give **three** nutrients needed by potato plants to grow healthily. *(3)*

    b As crops grow, they use up nutrients from the soil which then need to be replaced.

       i Give **one** way an intensive farmer would replace the nutrients. *(1)*

       ii Give **one** way an organic farmer would replace the nutrients. *(1)*

    c This article was published about the Colorado beetle in a farming newsletter.

> **Spuds in Trouble?**
>
> The Colorado beetle can destroy whole crops of potatoes and has destroyed many crops in the USA. Farmers are afraid that if this insect makes its way into the UK, by travelling on planes or ships, it could cause the same kind of destruction to their potato crops in this country.

       i Give **one** method used by an intensive farmer to protect his potato crops from the Colorado beetle. *(1)*

       ii Give **one** method used by an organic farmer to protect his potato crops from the Colorado beetle. *(1)*

       iii The newsletter goes on to advise farmers to use organic methods to deal with the Colorado beetle. Suggest **one** reason why. *(1)*

    d Why do some people prefer to buy organic potatoes instead of intensively farmed potatoes? *(1)*

    e Suggest why organic potatoes are more expensive than potatoes grown using intensive farming. *(1)*

2 Agricultural scientists research ways to improve crop yield.

    a Plants need to photosynthesise in order to produce food.

    Complete the word equation for photosynthesis reaction.

    carbon dioxide + ... $\longrightarrow$ .... + ... *(2)*

    b Plants also need nutrients from the soil. Agricultural scientists produce artificial fertilisers to provide these nutrients.

       i Name the type of nutrient plants need for healthy leaf growth *(1)*

       ii Name the type of nutrient plants need for good root development. *(1)*

    c *In this question you will be assessed on using good English, organising information clearly and using specialist terms as appropriate.*

    An agricultural scientist has produced three different fertilisers. She wanted to determine the best fertiliser for growing tomatoes. Describe a method she could use to do this. *(6)*

3 Farmers can choose from different farming techniques when rearing pigs for meat.

    a Name the type of farming shown in this photograph. *(1)*

b Give **two** ways in which farmers control the pigs' environment in this type of farming and explain how this helps increase the yield. (4)

c Suggest why animal rights groups may object to this type of farming for pigs. (1)

4 Farmers need to replace nutrients used up by growing crops. Ammonium nitrate is used in artificial fertiliser production, and is made by reacting ammonia with nitric acid. (1)

a Complete this word equation showing the production of ammonia.

.... + ... $\rightleftharpoons$ ammonia (2)

b Name the type of reaction represented by this symbol $\rightleftharpoons$ (1)

c Explain how increasing the temperature increases the rate at which this reaction occurs. (3)

d An industrial chemist calculated that 34 000 tonnes of ammonia could be made in theory from the starting materials available in a factory. However, she found that only 5100 tonnes were actually made in the conditions chosen. Calculate the percentage yield of ammonia. (2)

e Give **two** ways to increase the yield of this reaction. [H] (2)

5 A food health and safety inspector visits restaurants to check that they are complying with food health and safety regulations.

a i Restaurant owners must ensure that their customers cannot get food poisoning from the food in the restaurant. Choose the bacteria that could cause food poisoning.

Lactobacillus   Salmonella   Streptococcus (1)

ii Explain why a restaurant worker has to wear gloves when working in the kitchen. (2)

iii Complete the following questions about kitchen hygiene.

Kitchen surfaces must be cleaned with ....

Dishes must be cleaned with .... (2)

b Explain **two** ways that food can be treated with to ensure that growth of bacteria is slowed down or stopped. (4)

c *In this question you will be assessed on using good English, organising information clearly and using specialist terms where appropriate.*

The food health and safety inspector has to test some of the food to ensure it does not contain dangerous bacteria. Describe a method the food health and safety inspector could use to check for bacteria present in the food, including any safety measures he must carry out. (6)

AQA Examiner's tips

Question 5b has four marks. You need to give two examples and two explanations. You will get a mark for each example up to a maximum of two. You will then get a mark for each explanation upto a maximum of two. You will only get the explanation marks however, if they correctly link to the examples given.

AQA Examiner's tips

Make sure you read all the instructions carefully. The examiner has given lots of guidance for you to answer the question. Make sure you follow all the instructions. Again there are 6 marks so you should give a detailed scientific answer. You will be marked on your written skills so read through your answer at the end to check it makes sense and that your grammar and spellings are correct.

Making connections

In this chapter you will learn about how analytical scientists provide evidence that answers questions and solves problems. You will look at the important work that analytical scientists do as well as some of the techniques they use.

Figure 2 Testing the quality of water

Figure 3 A medical health technician testing urine samples

Figure 7 Forensic drug analysis

Figure 1 Analytical scientists carrying out laboratory work

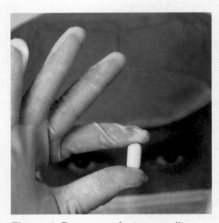

Figure 4 Drug manufacture quality control

Figure 6 DNA – our genetic fingerprint

Figure 5 A scientist undertaking chemical analysis

Identification of substances

Analytical scientists work in many sectors of society – from the manufacturing and pharmaceutical industries, to healthcare, forensic services and public protection. They must carry out their tests methodically and interpret their results accurately. If they don't, both their safety and that of the general public will be at risk.

In this chapter, you will explore many laboratory procedures that are found in the workplace. These include tests to identify chemicals, such as flame tests and titrations, and tests to identify biological material such as **DNA**.

a Notice the peaks on the three graphs in Figure 8 – red, black and green. Explain whether or not any of these peaks match.

b Suggest how DNA profiling like this could prove whether a man is or is not the father of two children.

The chemist in Figure 9 is holding samples of river water and tap water. She has added a chemical to each test tube. The yellow colour shows that the river water sample is contaminated, while the colourless tap water sample is safe to drink.

c Suggest an advantage of tests where the results may show a distinct colour change.

d State a colour change test that you have seen previously in science.

Scientists often use computers to help them with complex analysis.

e Suggest an example from the workplace where using computers eliminate the need for complex mathematics.

f Suggest a benefit of computer control in a scientific industry.

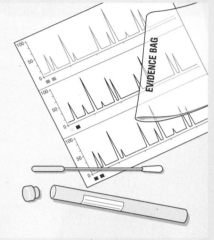

Figure 8 DNA sample and results

∞ links

For more information on DNA see 5.11 Blood and DNA and 5.12 DNA profiling.

Figure 9 Testing samples of river water and tap water

Figure 10 A scientist in a laboratory using instrumental analysis

Scientists @ work

Use the internet to research the work of an analytical scientist. The following careers database is a useful starting point:
www.connexions-direct.com/jobs4u
Follow the link to 'Explore Science and Maths'.

5.1 Introduction to analytical science (k)

Learning objectives

- Why is the work of analytical scientists important?
- What are the daily tasks of analytical scientists?

Analytical scientists

Examples of the work of analytical scientists include:

- **Quality control**, such as monitoring the production of foods, drinks, cosmetics and pesticides.
- **Research and development**, such as improving the quality of foodstuffs and developing new drugs.
- **Identification**, such as analysing body tissues and fluids to help diagnose diseases.
- **Analysing** materials found at crime scenes to assist in criminal investigations.

In their work, analytical scientists need to write clear reports, giving evidence of their results that others can check, reproduce for themselves and use. They may also refer to secondary sources (**reference literature**) to check the reproducibility of their results against those obtained by other scientists.

> **a** Suggest a disease that could be identified from a blood test.

Forensic scientists

In their work, forensic scientists help the police to investigate and detect crimes, convict offenders and free the innocent. Examples of evidence left by lawbreakers at a crime scene include: fingerprints, fibres from clothes, glass fragments, chemical compounds, blood and DNA evidence.

Forensic work begins by carefully observing the crime scene and accurately recording the materials found there. Only by collecting and storing the samples properly (without risk of cross-contamination) will a court of law accept the evidence as valid or trustworthy.

> **b** Look at Figure 1. When collecting the cigarette butt, how could an inappropriate technique lead to uncertainty about the validity, repeatability and reproducibility of the evidence?

Environmental protection

Most people agree that we need to protect the environment for future generations. Some people argue that to achieve this we must do things such as support the organic farming industry and protect the variety of life on Earth.

Defra, the **Department for Environment, Food and Rural Affairs**, helps the government to make policies and write laws covering environmental issues. Their work is based on scientific research and analysis. Environmental scientists working for Defra advise the government on issues such as the effects of climate change, and diseases like bird flu, on farming.

> **c** There is evidence that bird flu can be caught by humans. Explain the problems that could arise from wild birds or intensively reared chickens in the UK catching bird flu.

Figure 1 A forensic scientist, using a pair of tweezers, collects a cigarette butt found at a crime scene. The gloves and protective clothing help to prevent contamination of the evidence. He will analyse the cigarette butt for traces of skin cells and saliva. DNA profiling may then link a suspect to the crime scene.

Figure 2 Foot and mouth is a highly contagious disease affecting cloven-hoof animals. The only way to control foot and mouth disease is to cull the animals and stop people moving in and out of exclusion zones.

Healthcare and pharmaceuticals

Threats to your health include infectious diseases and environmental hazards, such as chemical poisons and radiation. The **Health Protection Agency** responds to health hazards and emergencies by providing advice and information. The Agency advises not only the general public, but also doctors and nurses, and the government. Their public health laboratory scientists are specialists in laboratory skills. We need their expertise to solve the complicated problems affecting our health.

About half of the medicines sold in the UK were developed in British laboratories. More than 90 companies make up the UK pharmaceutical industries. They employ over 73 000 people to develop, produce and market medical drugs.

Figure 3 Washing hands is the simplest way of preventing illnesses such as food poisoning and flu

Activity

Discuss the significance of these two facts:

- About 80 per cent of humans' food supply comes from just 20 kinds of plants.
- About one-eighth of known plant species are threatened with extinction.

Scientists @ work

Visit the Association of the British Pharmaceutical Industry (ABPI) schools' website at **www.abpischools.org.uk** to explore:

At Work With Science: How science is used in the pharmaceutical industry.

Students: Which animals are involved in creating new medicines.

14–16: Applied Science for a range of topics from Biotechnology to Stem Cell Research.

Activity

Carry out some research on how we should respond to the risks of either:

- infectious diseases caused by the HIV virus and *E. coli* bacteria
- a terrorist attack with chemical and biological weapons.

d Getting a medicine from the laboratory to the pharmacy involves:

1 Research and discovery by scientists targeting a particular illness.
2 Applying for patents, in case a rival company has already discovered it.
3 Legally required animal testing to check the active compounds for safety.
4 Human testing to identify its benefits and side effects, so the medicine can be registered.
5 Manufacture, first on a small scale before mass production.
6 Sales and marketing.
7 Continued monitoring of patients' reactions to the medicine after it has been passed for use with the general public.

Comment on the need for so many safety checks (3, 4 and 7 above).

Figure 4 Pharmaceutical industry laboratory

Summary questions

1 Suggest why monitoring the quality of foods and drinks is vital.

2 Why do manufacturers have research and development departments?

3 We use forensic science for other purposes besides helping the police investigate crimes. Suggest an example of the use of forensic methods when:
 a studying archaeological specimens
 b investigating the cause of an industrial accident.

4 Employees in pharmaceutical companies perform many tasks and use cutting-edge science to provide us with the medicines of the future. Suggest what skills a pharmaceutical scientist needs.

Key points

- Analytical scientists work in forensics, environmental protection and healthcare, which affect our wellbeing.

- Daily scientific tasks involve quality control, research and development and identification of materials.

5.2 Distinguishing different chemicals

Learning objectives

- How do analytical scientists identify substances?
- How can we use the structure of ionic compounds to explain their high melting points?
- Why do many covalent compounds have low melting and boiling points?
- What is the formula of certain ionic and covalent compounds?

Analytical scientists carry out tests on substances to use as evidence. They can identify a substance for evidence by looking at its melting point, boiling point and how it behaves when dissolved in water.

Ionic bonding and **covalent bonding** are two different ways that atoms can join to each other chemically. A compound with ionic bonds has a high melting point and usually dissolves in water. A compound with covalent bonds usually has a low melting point and is often insoluble in water.

Ionic bonding

An **ion** is an atom that has gained or lost an electron. **Electrons** are the tiny negative particles orbiting in an **atom**. An ionic bond occurs when:

- metal atoms lose an electron and become positive ions
- non-metal atoms gain electrons and become negative ions
- vast numbers of these oppositely charged ions attract and bond together in giant structures.

The small spheres in Figure 1 represent positively charged sodium ions. The large spheres are the negatively charged chloride ions. **Electrostatic attraction** between neighbouring ions, operating in all directions, holds the giant lattice together. This makes the structure strong. To break the strong bonds between so many positive and negative ions requires a lot of energy. This explains why **ionic compounds**, like sodium chloride (common salt), have high melting points.

In ionic compounds:

- Group 1 metals, like sodium (Na), form 1+ ions.
- Group 2 metals, like magnesium (Mg), form 2+ ions.
- Group 6 non-metals, like sulfur (S), form 2− ions.
- Group 7 non-metals, like chlorine (Cl), form 1− ions.

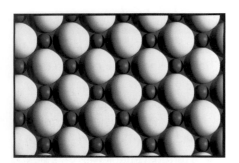

Figure 1 The closely packed ions in the giant structure of sodium chloride

Group 1 – alkali metals (shaded yellow)

Group 2 – alkaline earth metals

Group 7 – halogens

Group 6

H							He
Li	Be	B	C	N	O	F	Ne
Na	Mg	Al	Si	P	S	Cl	Ar
K	Ca						

Figure 2 The first 20 elements of the periodic table

Metal ions		Non-metals ions	
After losing 1 electron:		*After gaining 1 electron:*	
sodium	Na^+	chloride	Cl^-
potassium	K^+	bromide	Br^-
silver	Ag^+	nitrate	NO_3^-
		hydroxide	OH^-
After losing 2 electrons:		*After gaining 2 electrons:*	
magnesium	Mg^{2+}	oxide	O^{2-}
calcium	Ca^{2+}	sulfide	S^{2-}
iron(II)	Fe^{2+}	sulfate	SO_4^{2-}
copper(II)	Cu^{2+}	carbonate	CO_3^{2-}
zinc	Zn^{2+}		
lead	Pb^{2+}		

Table 1 Metal and non-metal ions. Notice that some negative ions, like sulfate, are made of groups of non-metal atoms.

Writing formulae for simple ionic compounds

Ionic compounds have no overall charge. So to work out a formula, **match the charges** so that the positive and negative charges cancel each other out. You may need different numbers of each type of ion. Notice that you write the formula without the charges.

e.g.
sodium chloride	NaCl	(1+, 1–)
magnesium oxide	MgO	(2+, 2–)
sodium sulfate	Na_2SO_4	(2+, 2–)

(We need 2 Na^+ to cancel the charge on SO_4^{2-}.)

copper(II) nitrate $Cu(NO_3)_2$ (2+, 2–)

(The (II) shows Cu(II) has a 2+ charge.)

a What is the formula of copper(II) bromide?

Covalent bonding

When non-metal atoms join together they form covalent bonds. They do this by sharing pairs of electrons. This creates strong bonds between the atoms **within** each **molecule**, e.g. oxygen O_2, water H_2O, ethanol C_2H_5OH and glucose $C_6H_{12}O_6$.

Yet the forces of attraction **between** the molecules are weak. **Covalent compounds** have low melting points and boiling points. That is because you only need a little energy to separate the molecules from each other.

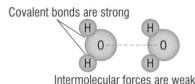

Covalent bonds are strong

Intermolecular forces are weak

Figure 5 Weak forces between covalent molecules

b Why is carbon dioxide (CO_2) a gas at room temperature?

Many substances obtained from living materials are **organic** compounds with covalent bonding, and all contain the element carbon. We produce ethanol by fermenting sugar with yeast for alcoholic drinks. We also use ethanol for fuel and as a **solvent** in industry and in laboratories.

Summary questions

1 Why do ions attract in ionic bonding?

2 Name the type of bonding in a compound which has a giant lattice, held together by strong electrostatic forces of attraction.

3 Name the type of bonding in an organic compound.

4 Which is more likely to dissolve in water, an ionic compound or a covalent compound?

5 Explain why common salt (sodium chloride) has a high melting point, but water has a low melting point.

6 What is the formula of each of these molecules?

a water	**b** carbon dioxide	**c** calcium oxide	**d** zinc chloride
e lead sulfate	**f** glucose	**g** ethanol	**h** iron(II) nitrate

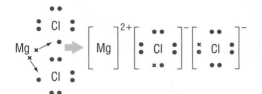

Figure 3 The ionic bonding of magnesium chloride ($MgCl_2$). The metal atom loses electrons, the non-metal atoms gain them.

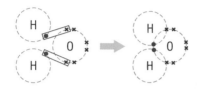

Figure 4 The covalent bonding of water (H_2O) – the atoms share the electrons

⚲ links

For further information on the test for: ethanol (C_2H_5OH), see 5.4 Breathalysers; glucose ($C_6H_{12}O_6$), 2.12 Standard procedures for maintaining health and fitness.

Key points

- The properties of substances can be used to help identify the substances.

- Ionic compounds have ionic bonds between positive metal ions and negative non-metal ions, as in sodium chloride NaCl. The forces of attraction operate in all directions. It takes a lot of energy to break so many strong bonds and separate the ions. Therefore, ionic compounds have high melting points.

- Most covalent compounds have low melting points and boiling points. Organic compounds, such as C_2H_5OH and glucose $C_6H_{12}O_6$, have covalent bonds. To separate their molecules from each other only needs a little energy.

5.3

Testing for ions

Learning objectives

- How are chemicals analysed to test for ions?

- What are precipitation reactions and how are they used?

Figure 1 A technician using a mortar and pestle to grind a soil sample. She will use standard tests to analyse for metal ion contamination.

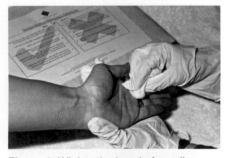

Figure 2 Wiping the hand of a police suspect. The swab will be analysed for the presence of chemicals.

Environmental technicians may have to test for ions contaminating a river. Forensic scientists often have to analyse chemical evidence found at the crime scene and on suspects' clothes and hands. These analytical scientists always aim for the most appropriate, yet simplest method to carry out a task. Time and money must not be wasted.

Practical

Analysing ionic compounds

Remember ionic compounds are formed between metals (the positive ions) and one or more non-metals (the negative ions).

The identity of a chemical may provide us with vital evidence. You have probably seen chemical reactions in which ions seem to 'swap partners'. A and C below are positive ions, and B and D are negative ions.

$$AB + CD \longrightarrow AD + CB$$

Here is an example:

lead nitrate + potassium iodide ⟶ lead iodide + potassium nitrate

$$Pb(NO_3)_2(aq) + 2KI(aq) \longrightarrow PbI_2(s) + 2KNO_3(aq)$$

The label (aq) means dissolved in water (or *aq*ueous), and (s) means a **s**olid.

Notice that solutions of lead nitrate and potassium iodide are both colourless. Now, wearing eye protection, mix them together in a test tube, one drop at a time.

- What happens?

The lead iodide forms as a **precipitate**. The lead iodide does not dissolve. We call this a **precipitation reaction**.

Precipitation reactions

In a precipitation reaction, a suspension of an insoluble solid forms when two solutions react together. The colour of the solid produced can give us an indication of one of the reactants we started with.

Scientists use precipitation reactions to analyse the contents of unknown solutions. The solution may come from a polluted river and the technician needs to provide information to help identify who is responsible. Two tests are needed, one to test for positive metal ions and another to test for negative non-metal ions.

a What is a precipitate?

Practical

The sodium hydroxide (NaOH) test for positive metal ions

1 Wearing chemical splash proof eye protection and taking care, put 2–3 cm³ of the solution being tested into a test tube.

2 Carefully add a few drops of 0.1 mol/dm³ sodium hydroxide solution (irritant).

Metal ion in solution	Observation after adding NaOH solution
Lead (Pb^{2+})	White precipitate, which dissolves if we add more NaOH
Calcium (Ca^{2+}),	White precipitate, which doesn't re-dissolve
Copper(II) (Cu^{2+})	Blue-green, jelly-like precipitate
Iron(II) (Fe^{2+})	Green-grey, jelly-like precipitate
Iron(III) (Fe^{3+})	Red-brown, jelly-like precipitate

In reactions with solutions of the ions above the precipitate is the metal hydroxide, e.g.

lead nitrate + sodium hydroxide ⟶ lead hydroxide + sodium nitrate

$$Pb(NO_3)_2(aq) + 2NaOH(aq) \longrightarrow PbOH_2(s) + 2NaNO_3(aq)$$

b What colour is copper hydroxide?

Practical

Tests for negative non-metal ions

The following table lists the tests for negative non-metal ions. Carry out each of these tests and note the observation.

Non-metal ion in solution	Test	Observation
Carbonate (CO_3^{2-})	Add dilute acid.	Carbon dioxide gas given off, which turns limewater cloudy
Chloride (Cl^-)	Add a few drops of dilute nitric acid *then* add a few drops of silver nitrate solution	White precipitate
Sulfate (SO_4^{2-})	Add a few drops of dilute hydrochloric acid *then* add a few drops of barium chloride solution	White precipitate

Summary questions

1 What is the name of the precipitate produced when solutions of copper(II) sulfate and sodium hydroxide react together?

2 What is the name of the compound dissolved in the following solution?
   ● When you add sodium hydroxide (NaOH) solution to it, a red-brown precipitate forms.
   ● When you add dilute nitric acid then silver nitrate solution, a white precipitate forms.

links

For information on how scientists distinguish between the metal ions sodium and potassium (Na^+ and K^+), see 5.5 Flame tests.

Key points

● We can use a precipitation reaction to identify ions in solution.

● A precipitate is an insoluble solid formed when mixing two solutions. The colour of a precipitate can help you identify a substance.

● The sodium hydroxide test (NaOH) uses precipitation reactions to identify metal ions:

AB + CD ⟶ AD + CB

solutions precipitate

5.4 Breathalysers

Learning objectives

- How can we test for ethanol?
- How has breathalyser technology advanced?

Testing for ethanol

The alcohol in alcoholic drinks is ethanol, C_2H_5OH (or CH_3-CH_2-OH). You absorb ethanol into your blood through your stomach wall. Most is broken down into carbon dioxide and water in your liver. The rest of the ethanol leaves your body in sweat, breath and urine. Breathalysers (i.e. *breath analysers*) work by testing for ethanol in your breath.

To test for ethanol you can use acidified potassium dichromate, which changes colour from orange to green. Robert Borkenstein used this chemical reaction to design the original breathalyser. He used photocells connected to a meter to monitor any colour change.

Figure 1 Alcohol test. You test for ethanol with acidified potassium dichromate, which changes from orange to green

Did you know ... ?

If you're convicted of drink-driving you will:

- lose your licence for at least 12 months
- face a maximum fine of £5000
- face up to six months in prison
- pay up to three times as much for car insurance.

Practical

Making your own breathalyser

1 Using tweezers, soak a piece of mineral wool in breathalyser solution, then poke it into a small pipette.

2 Add a few drops of ethanol to the balloon, blow it up and seal it with a Hoffman clip.

3 Mount the pipette in a clamp stand over a sink, attach the balloon and release the air slowly.

4 Repeat the experiment with another balloon filled only with air.

5 Rinse your mouth with alcohol-based mouthwash before blowing the balloon up and repeat the test again.

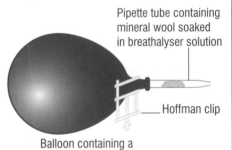

Balloon containing a few drops of ethanol

Figure 2 Simple dichromate breathalyser

Safety: Some people have a latex allergy. The breathalyser solution is toxic and an irritant. Ethanol is toxic and flammable. Wear eye protection and gloves. Wash your hands after the experiment.

Figure 3 A driver, suspected of drink-driving, breathing into the mouthpiece of a breathalyser. Even if he tests negative, he may still need to take a drugs test. Sadly, driving under the influence of drugs is an increasing hazard. 60% of drivers and passengers in fatal road crashes have drugs in their blood.

Look at the task in the practical box to make your own breathalyser using acidified potassium dichromate solution.

a What does the control test 4 prove?

b i Test 5 could result in a 'false positive'. How can a motorist fail a breath test without drinking?

ii Why would a blood or urine test prove this motorist's innocence?

Think! Drink-driving

If you **drive** at twice the current legal alcohol limit, you are at least 30 times more likely to cause a road crash than a driver who has not been drinking.

The only safe way is not to **drink** alcohol at all if you are **driving**.

These are the legal limits for driving in the UK:

- Breath: 35 mg/100 cm^3
- Blood: 80 mg/100 cm^3
- Urine: 107 mg/100 cm^3

The police carried out half a million breath tests in the UK last year. 100 000 of these proved positive. The latest figures also show that nearly 600 people were killed in crashes in which the driver was over the legal limit, and 2350 were seriously injured.

The most dangerous times of day are when children go to school and people return from work.

Young men aged 17–29 are the main casualties and drink-drive offenders.

Activity

'Don't turn your night out into a nightmare.'

Devise your own anti drink-drive slogan and poster.

Advances in breathalysers

In 1954, Borkenstein invented the breathalyser which uses an **ethanol–dichromate reaction** to produce a colour change.

Since the 1980s, in order to achieve greater accuracy and a faster response time, breathalysers now use either:

1 Infrared spectroscopy – detecting the concentration of ethanol by the amount of infrared radiation absorbed.

2 Fuel cell technology

If more ethanol is present, there is a greater current in the fuel cell.

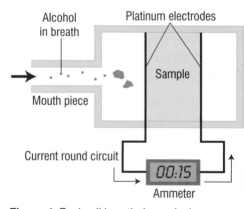

Figure 4 Fuel cell breathalyser design

c What three methods have scientists used to make breathalysers?

Summary questions

1 Copy and complete the following sentences:

The chemical reaction in the design of the original breathalyser is between .... potassium .... and .... – the colour changing from .... to .... if alcohol is present.

2 Why are accuracy and response time important in a breathalyser?

3 Compare modern breathalysers with Borkenstein's original design.

∞ links

For information on how to use infrared spectroscopy to detect drugs in the crime lab, see 5.10 Modern analytical instruments. **[H]**

AQA Examiner's tip

In an examination you may be asked to 'discuss the advances in breathalyser technology over time', but you do not need to remember all the details of the breathalyser design.

Key points

- You test for ethanol with acidified potassium dichromate, which changes colour from orange to green.

- Nowadays breathalysers are made using either infrared spectroscopy or fuel cell technology, rather than relying on the colour change in the acidified potassium dichromate and ethanol reaction.

5.5 Flame tests

Learning objectives

- What tests can we use to identify metal ions?

- How can we improve the accuracy and reproducibility of the evidence we collect?

Sodium Na⁺

Potassium K⁺

Calcium Ca²⁺

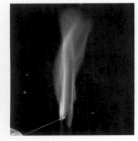

Copper Cu²⁺

Figure 1 Flame test colours of certain metal ions

Analytical scientists use flame tests to identify positive metal ions in samples of material. For example, a forensic scientist may recover a soil sample from a crime scene. Chemical analysis can determine which metal ions are present in the soil.

The compounds of some metals give a distinctive flame colour. Although a flame test may tell you what metal ion is present, it will not tell you how much of that ion is in a sample. Consequently, scientists usually carry out flame tests as well as other analytical techniques.

a In a flame test, what metal ion gives the following colour?

   i golden yellow   **ii** lilac   **iii** brick red   **iv** blue-green

b Why might a forensic scientist flame test the soil from the shoe of a suspect?

Practical

Standard procedure for a flame test

1 Wearing chemical splash proof eye protection throughout, clean the nichrome wire loop in concentrated hydrochloric acid. Tap to remove any excess. Take great care as the concentrated acid is corrosive.

2 Place the loop in the hot part of a Bunsen flame.

3 Repeat steps 1 and 2 until there is no change in flame colour (other than dull orange). The wire is now clean.

4 Dip the loop in hydrochloric acid again.

5 Dip the loop into the sample.

6 Place the loop in the edge of a hot Bunsen flame.

7 Record the colour of the flame.

8 Repeat steps 1 to 4 before testing another sample.

Safety: Be aware of the hazards before you try a flame test. Your safety matters. In your risk assessment, consider the risks of hot concentrated hydrochloric acid spitting into your eye or onto the table. See the CLEAPSS Hazcard 47.

Sarah wants to improve the accuracy of her flame test results. She suspects her wire loop is not clean enough.

- What should Sarah do to improve the accuracy of her test?

- Why would a contaminated wire loop affect the reproducibility of Sarah's results?

c Copy and complete the two tables below. Based on the tests outlined here and in 5.3, identify these two substances found on the site of a demolished factory with plans for new housing.

Specimen A

Test	Observation	Conclusion
A flame test was carried out on solid A	The flame colour was lilac	
A spatula of solid A was placed in a test tube of hydrochloric acid	The mixture fizzed. The gas given off turned limewater milky	
Specimen A is …		

Specimen B

Test	Observation	Conclusion
A flame test was carried out on solid B	The flame colour was blue-green	
Sodium hydroxide was added to a solution of B	A blue-green, jelly-like precipitate was formed	
A spatula of solid B was placed in a test tube of hydrochloric acid	No reaction	
A few drops of dilute nitric acid were added to a solution of B. Then silver nitrate solution was added	A white precipitate formed	
Specimen B is …		

Summary questions

1 Copy and complete:

We can use flame tests to find out which .... ions are in a sample, each producing a different flame ....

2 Write about a situation where the flame test of sea water could prove useful.

3 How well do you think a flame test might work on a mixture of salts containing different metal ions?

4 Your technician gives you 4 M (4 mol/dm^3) hydrochloric acid to clean your nichrome wire. It is difficult to remove sodium ions from your wire loop. Your partner suggests using some more concentrated hydrochloric acid. What do you think? Is there any alternative?

Practical

Spray bottle flame test for soluble salts

Safety: Only carry out this test under direct teacher supervision and be aware of the hazards. Wear eye protection and always direct the spray away from yourself and your audience.

1 Use small spray bottles containing concentrated aqueous chloride solutions.
2 Spray each one in turn into a Bunsen flame.
3 Identify the metal chloride present.

links

For further tests on ions look back at 5.3 Testing for ions.

Key points

- You can use a flame test to identify a metal ion in a solid:

Metal ion	Flame colour
Sodium	Bright yellow
Potassium	Lilac
Copper	Green-blue
Calcium	Brick red

- A flame test is not accurate or reproducible if the wire loop is contaminated with another chemical, hence the need to clean it thoroughly in concentrated hydrochloric acid.

5.6

Balanced equations

Learning objectives

- What are relative atomic mass, relative formula mass and a mole?

- How do you write balanced equations? [H]

- How can we calculate the masses of reactants and products?

Figure 1 A mole of a substance is its relative formula mass in grams

AQA Examiner's tip

On a periodic table, the relative atomic mass is beside the symbol.

To calculate the relative formula mass, use the formula of the compound and add together the relative atomic masses of atoms it contains.

For example, the relative formula mass of sulfuric acid, H_2SO_4, is 98.

$(1 \times 2) + 32 + (16 \times 4)$
$= 2 + 32 + 64 = 98$

Moles

When atoms bond chemically, molecules can be formed. The formula for an oxygen molecule is O_2. The subscript 2 tells you there are two atoms of oxygen in an oxygen molecule. The **relative atomic mass** of oxygen is 16. Think of an oxygen atom as 16 times heavier than the lightest of all atoms, the hydrogen atom. Hydrogen has a relative atomic mass of 1.

Therefore the **relative formula mass** of O_2 is 32 as $16 \times 2 = 32$.

Similarly the relative formula mass of water, H_2O, is 18 as $(1 \times 2) + 16 = 18$.

> **a** What is the relative formula mass of:
>
> **i** methane, CH_4 **ii** sodium carbonate, Na_2CO_3?
>
> (The relative atomic mass of C = 12 and Na = 23.)

32 g of O_2 (oxygen) and 18 g of H_2O (water) both contain exactly the same number of molecules. That amount of substance is called a 'mole'.

If you work out the relative formula mass of a substance and weigh out that mass in grams, it actually contains 6.02×10^{23} molecules!

Chemical equations

Figure 2 shows the reactants and products when methane burns in oxygen.

   Combustion

O_2 $\quad$ CH_4 $\quad$ O_2 $\qquad\qquad$ CO_2 $\quad$ H_2O

Figure 2 Burning methane with a Bunsen burner

The 'word equation' for this reaction is:

methane + oxygen $\longrightarrow$ carbon dioxide + water

The next equation is a 'symbol equation', but it is not 'balanced':

$$CH_4 + O_2 \longrightarrow CO_2 + H_2O$$

Elements	Reactants (on left)	Products (on right)
carbon, C	1	1
hydrogen, H	4	2
oxygen, O	2	3

You balance an equation by putting numbers in front of the formulas:

$$CH_4 + 2O_2 \longrightarrow CO_2 + 2H_2O$$

The table below shows the number of atoms in the balanced equation

Elements	Reactants (on left)	Products (on right)
carbon, C	1	1
hydrogen, H	4	4
oxygen, O	4	4

Higher

b Balance the following equations:

   i $Ca + O_2 \longrightarrow CaO$

   ii $H_2 + Br_2 \longrightarrow HBr$

   iii $Al + I_2 \longrightarrow AlI_3$

In a balanced equation there is the same number of atoms on both sides.

??? Did you know …?

You put NaCl and CH_3COOH on your fish and chips. That's sodium chloride and ethanoic acid. They are not poisons, just salt and vinegar!

Equations and calculations

Calculating the relative formula mass (in grams) for each symbol gives:

$$CH_4 \quad + \quad 2O_2 \quad \longrightarrow \quad CO_2 \quad + \quad 2H_2O$$

$$12 + (1 \times 4) \quad 2(16 \times 2) \quad\quad 12 + (16 \times 2) \quad 2(1 \times 2 + 16)$$

$$= 16\,g \quad\quad = 64\,g \quad\quad\quad = 44\,g \quad\quad\quad = 36\,g$$

1 mole of CH_4 + 2 moles of O_2 produce 1 mole of CO_2 + 2 moles of H_2O.

Burning 16 g of methane produces 44 g of carbon dioxide.

Notice that the total mass of the reactants equals the total mass of the products. There are 80 g on both sides of the equation.

c This equation shows the neutralisation reaction between nitric acid and calcium carbonate:

$$?HNO_3 + CaCO_3 \longrightarrow Ca(NO_3)_2 + H_2O + CO_2$$

   i When balanced with **$2HNO_3$**, how many atoms of oxygen, nitrogen and hydrogen are on each side?

   ii If 100 g of calcium carbonate react, how many grams of carbon dioxide are produced?

   iii How many grams of nitric acid do you need to react with 100 g of calcium carbonate?

Relative atomic masses: H=1, C=12, N=14, O=16, Ca=40

Summary questions

1 Calculate the relative formula mass for:
  a nitric acid, HNO_3, **b** calcium carbonate, $CaCO_3$, **c** calcium nitrate, $Ca(NO_3)_2$

  (Relative atomic masses: H = 1, C = 12, N = 14, O = 16, Ca = 40)

2 How many moles of oxygen are there in 64 g?

3 Drain cleaner is strongly alkaline, containing sodium hydroxide. Trading Standards scientists can check its concentration by neutralising it with sulfuric acid. The equation for the reaction is:

$$2NaOH + H_2SO_4 \longrightarrow Na_2SO_4 + 2H_2O$$

  a Show that the equation is balanced. [H]

  b How many atoms of oxygen are on each side of the equation?

  c What mass of acid neutralises 80 g of sodium hydroxide?

  d What mass of acid would neutralise 40 g of sodium hydroxide?

(Relative atomic masses: Na = 23, S = 32)

Key points

● The relative atomic mass of an element is the mass of its atoms compared to the mass of other atoms. The relative formula mass in grams is known as a mole of that substance.

● To write a balanced equation put numbers in front of the formulae to ensure each element has the same number of atoms on both sides of the equation. [H]

● To calculate the masses of reactants and products, calculate the relative formula masses of each substance in a balanced equation.

The use of science in analysis and detection

5.7 Titrations

Learning objectives

- How do you carry out a titration?
- How can you improve the accuracy and reproducibility of results from titrations?

Standard procedure

A **titration** is an important standard procedure used by analytical scientists. It is the established method to determine the concentration of a solution. This could involve finding the amount of acid in rain water, lactic acid in milk, ethanoic acid in vinegar, or metal ions in polluted river water.

Titration requires care, coordination and concentration. You need to carry out a risk assessment and practise the techniques before starting.

Practical

Acid–base titration

Aim: To find the volume of hydrochloric acid needed to neutralise 20 cm³ of sodium hydroxide solution.

The acid is a **standard solution** with a concentration of 0.4 mol/dm³ – meaning 0.4 moles of HCl are dissolved in 1 litre of hydrochloric acid solution. 1 litre = 1000 cm³ = 1 dm³

1. Using a pipette (and safety filler) transfer 20 cm³ of sodium hydroxide solution into a conical flask.
2. Add a few drops of screened methyl orange indicator, and put a white tile under the flask.
3. Fill a burette using a funnel with your standard solution of 0.4 mol/dm³ hydrochloric acid. Drain some acid into a beaker to remove the air below the tap. Record the starting volume to the nearest 0.05 cm³.
4. Slowly add the acid (1 cm³ at a time) to the flask of sodium hydroxide solution. Swirl the flask and stop at the **end point**, when the indicator just changes colour, from green to violet. The solution is now neutral. Record the final volume. Calculate the volume of acid used. This is your trial or rough experiment.

5. Repeat the titration more carefully, swirling and adding acid drop by drop near the end point. Continue repeats until you get two accurate volumes within 0.1 cm³. Average these two accurate titration volumes. Call this volume V cm³.

Safety: Wear eye protection. The sodium hydroxide solution is an irritant alkali, with the concentration less than 0.5 mol/dm³. The dilute hydrochloric acid is a low hazard. The screened methyl orange indicator is toxic if swallowed.

Figure 1 Swirl and add acid drop by drop near the end point

So how many moles of acid react with the alkali?

Suppose your volume of acid, V = 22.5 cm³ neutralised the 20 cm³ of NaOH. Remember the acid contained 0.4 moles of HCl in 1000 cm³ of solution. So in 1 cm³ of acid we have 0.4 ÷ 1000 moles of acid.

i   Therefore in 22.5 cm³ we have (0.4 ÷ 1000) x 22.5 moles of acid.

ii   = 0.009 moles of hydrochloric acid.

The equation for this neutralisation reaction is:

$$HCl(aq) + NaOH(aq) \longrightarrow NaCl(aq) + H_2O(l)$$

(Remember (aq) means dissolved in water and (l) means liquid.)

Since 1 mole of HCl neutralises 1 mole of NaOH, 0.009 moles of HCl neutralise 0.009 moles of NaOH.

a What is the relative formula mass of sodium hydroxide NaOH?

(Relative atomic masses: H = 1, O = 16, Na = 23)

b 1 mole of NaOH contains 40 g. What is the mass of 0.009 moles of NaOH?

c If 0.009 moles of NaOH are dissolved in 20 cm³, how many moles are dissolved in every 1 cm³?

d Multiplying 0.00045 by 1000 gives you the concentration of the sodium hydroxide solution. What is the concentration of the solution in mol/dm³?

Microscale titration

Analytical scientists often work with small quantities of materials. This not only saves money but reduces hazards in the event of spills. In microscale titration, a 2 cm³ pipette linked to a 10 cm³ syringe replaces the normal burette.

Practical

Microscale titration

Aim: To find the volume of 0.1 mol/dm³ hydrochloric acid needed to neutralise 1.0 cm³ of sodium hydroxide solution (about 0.1 mol/dm³) using phenolphthalein indicator.

- Using a pipette (and safety filler) transfer 1.0 cm³ of sodium hydroxide solution into a clean 10 cm³ beaker.

- Add one drop (no more) of phenolphthalein indicator.

- Fill the syringed pipette with 0.1 M hydrochloric acid, avoiding air bubbles. Clamp firmly and record the starting volume.

- Titrate the acid into the alkali by pressing very gently on the syringe. Swirl and stop when the indicator just turns from pink to colourless. Calculate the volume of acid added at the end point.

- Repeat the titration until you get reproducible measurements. Call this volume V cm³.

Safety: As before. Wear eye protection. Phenolphthalein indicator solution is highly flammable.

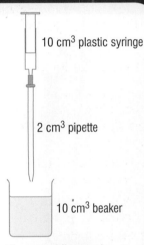

Figure 2 Microscale titration apparatus

10 cm³ plastic syringe

2 cm³ pipette

10 cm³ beaker

Suppose 1.0 cm³ of sodium hydroxide neutralises 1.25 cm³ of hydrochloric acid (of concentration 0.1 mol/dm³). If 0.1 moles of HCl are present in 1000 cm³ of hydrochloric acid, then the number of moles in 1.25 cm³ = (1.25 ÷ 1000) × 0.1 = 0.000 125 moles

There must also be 0.000 125 moles of NaOH in the sodium hydroxide, as it neutralises the acid. The concentration of NaOH is 0.000 125 moles in 1.0 cm³, so in 1000 cm³ there is 0.000 125 × 1000 = 0.125 moles = 0.125 mol/dm³.

As there are 40 g of NaOH in 1 mole of sodium hydroxide, i.e. (23 + 16 + 1) then the number of grams of NaOH in 0.000 125 moles = 40 × 0.000 125 = 0.005 g.

Summary questions

1 Draw a results table for a titration.

2 20.0 cm³ of hydrochloric acid reacted with 25 cm³ of 0.1 mol/dm³ sodium hydroxide.
   a Calculate the moles of alkali (and acid used).
   b The balanced equation for the reaction is:
   $HCl + NaOH \longrightarrow NaCl + H_2O$
   How many moles of hydrochloric acid reacted with the 25 cm³ of 0.1 mol/dm³ sodium hydroxide?
   c Calculate the concentration of the acid (in mol/dm³).

Key points

- In a titration you add acid from a burette to a pipette-measured volume of reactant. An indicator shows the end point.

- Taking care, repeat with clean apparatus until results are repeatable and reproducible to ensure accuracy.

5.8

Chromatography (k)

Learning objectives

- What is chromatography and how does it work?
- What is the difference between paper and thin-layer chromatography?
- How do we identify the substances present?

Chromatography can be used to separate and identify small quantities of chemicals in a mixture. You can also use chromatography to detect the number of components in a mixture. It is ideal for separating out the dyes found in inks. Therefore forensic scientists use the technique in forgery cases when a document might have been altered.

In the pharmaceutical and petrochemical industries, analytical scientists use chromatography to check the purity of drugs and fuel.

Look at the chromatograms from the ink in a poison pen letter and from some suspects' pens in Figure 1:

a Which suspect wrote the letter: A, B, C, D, E or F?

b Why is this test not a proof of guilt?

Types of chromatography

There are several types of chromatography. At the simpler end there is **paper chromatography** and **thin-layer chromatography (TLC)**.

The mixture you are testing, the **solute**, dissolves in the solvent. As the solvent moves through the **stationary substance** (the paper or thin-layer medium), different colours in the mixture are carried different distances. The different chemicals in the mixture separate at different rates. Every type of chromatography involves a **mobile solvent** passing through a stationary substance. Scientists describe the medium the solvent passes through as the **stationary phase**, while the solvent is the **mobile phase**.

Figure 1a Ink from the letter

Chromatography depends on the relative attractions of the molecules to the mobile solvent and the stationary substance. Some cling (stick) more strongly to the stationary substance. These are stopped first. Chemicals that do not bond as easily to the stationary substance move on by. In addition, substances that are more soluble in the solvent travel fastest through the stationary substance.

Higher

You cannot always use water as the solvent. The mixture may not dissolve in water. Ethanol is a non-aqueous solvent that could dissolve biro ink.

In paper chromatography, the stationary substance is the paper containing trapped water molecules. Paper contains about 10% water.

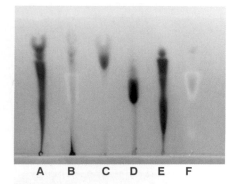

A B C D E F

Figure 1b Chromatograms from the suspects' pens

In thin-layer chromatography, the stationary substance is a tiny layer of powder. The powder is coated on to a plastic sheet or a glass plate as a paste and left to dry. Chemicals moving past attach to the powder to different extents.

Liquid–gas chromatography

∞ links

For more information on gas–liquid chromatography, see 5.10 Modern analytical instruments.
For information on DNA profiling and electrophoresis, see 5.12 DNA profiling.

Chemicals from drugs fix into hair as it grows. The forensic scientist cuts the hair into 1 cm lengths. Then a robot injects these, one at a time, into a gas chromatography machine. The sequence of hairs reveals the drug user's habits going back over the past few weeks. In gas chromatography you separate vapours from a volatile liquid, rather than separating the mixture within the liquid itself.

Maths skills

The **retention factor** equation:

$$R_f = \frac{\text{distance travelled by substance}}{\text{distance travelled by solvent}}$$

So, if the total distance a solvent travels through the medium is 10 cm, and the mean distance a chemical travels is 5 cm, then the retention factor for that chemical $R_f = 5 \div 10 = \mathbf{0.5}$.

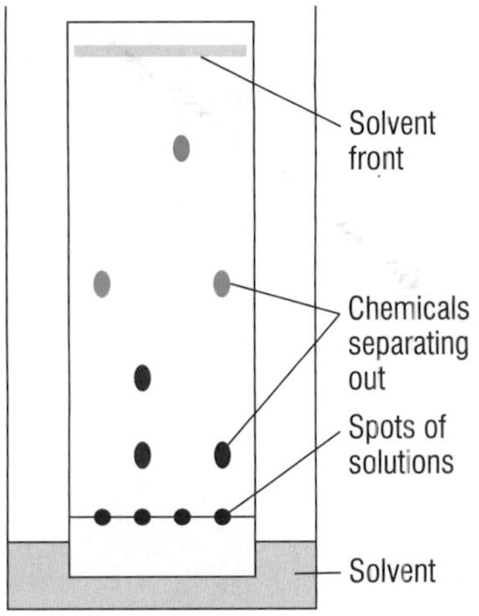

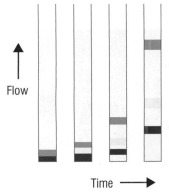

Figure 2 Chromatography

Practical

Making a chromatogram

- Dissolve your samples in a suitable solvent. Make sure the solution is concentrated. (If the solvent is flammable ensure there are no naked flames.)
- Place a spot of each sample on a pencil line 2 cm from the bottom of the paper (or TLC plate).
- Add 1 cm of solvent to your container and stand your paper in the solvent.
- Cover the container and leave until the solvent rises up **almost** to the top of the paper.
- Remove the chromatogram, mark the height the solvent rose to (called the **solvent front**), and leave to dry.
- For each colour calculate the retention factor, R_f.

Figure 3 How a chromatogram changes with time

Flow

Time →

c Why are different coloured chemicals in the mixture carried different distances by the solvent?

d How does a taller container in chromatography help?

Summary questions

1 All the substances (A, B, C, D) shown in Figure 4 are pure and not mixtures.
   a How do we know?
   b What does the distance X measure?
   c Spot Y is smallest. Is this a problem? Explain your answer.
   d Could substances A and D be the same chemical? Explain your answer.

2 Give an example of the use of chromatography in industry.

3 **a** A police suspect's hands are covered in paint. How could you match this paint to some graffiti?
   b Why might a forensic scientist decide to view a chromatogram under ultraviolet light?

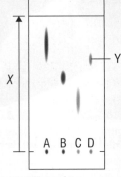

Figure 4

Key points

- Chromatography is used to separate and compare mixtures such as inks.

- In paper chromatography the stationary phase is the paper containing trapped water molecules. Paper contains about 10% water. In thin-layer chromatography (TLC) the stationary phase is a layer of powder coated onto a sheet.

- Different substances travel different distances depending on how each:
  - is attracted to the stationary phase
  - travels in the mobile phase (solubility in the solvent).

5.9

Microscopic evidence

Learning objectives

- Why are the instruments of analytical scientists likely to give more accurate results than those in a school laboratory?
- What are the distinctive features of fibres, soil, seeds, pollen grains and layers of paint?

Many types of evidence are far too small to see with the naked eye. Analytical scientists use standard light microscopes, such as we have in schools and colleges. They also use more expensive and powerful microscopes. These give better **resolution** and **contrast**.

Resolution

The better the resolution the sharper the image.

With a **scanning electron microscope** (SEM) we can see objects as small as 10nm (10 **nanometres**). 1nm is one millionth of one millimetre, as small as many molecules. SEMs give a 20 × better magnification than the best light microscopes.

Contrast

The better the contrast the clearer the image.

We can improve contrast by using:

- A **stain** – A coloured dye can bring out hidden detail.
- A **comparison microscope** – This is actually two microscopes side-by-side. We then compare the two images, on a computer screen, to see if they match.

> **a** Look at Figure 2. What does it prove if these scratch marks match?
> **b** Does this prove that a suspect fired the bullet found at the crime scene? Explain.
> **c** What other evidence could prove the supect fired the gun?

- A **polarising microscope** – You may know that wearing Polaroid glasses reduces glare. Placing paint pigments and fibres between crossed-Polaroids (at 90° to each other) can brighten the object while darkening the background.

The resolution and contrast of these microscopes provide more precise and reliable evidence than the microscopes used in schools and colleges.

Figure 1 A technician holds the scanning electron microscope (SEM) image of a moss, found on the clothes of a suspect

Figure 2 Used bullet cartridges in two microscope holders for comparison. Unique scratch marks occur when a bullet moves down the barrel and when a firing pin hits the back of a cartridge. Scientists compare marks on shells recovered from a crime scene, to marks on a bullet test-fired from a suspect's gun.

Figure 3 Collecting paint evidence. Layers show a car has recently been re-sprayed. The paint is compared with known paint samples.

> **d** Invent a story based on the three photographs in Figure 3.

Trace evidence

Trace evidence often links a suspect to the scene of a crime. Remember, 'every contact leaves a trace'. Paint flecks transfer when burglars break and enter, vandals scribble graffiti, and cars crash in road traffic accidents. Paint flecks and glass fragments, dust and soil, fibres and hairs, seeds and pollen, can all transfer to and from the clothing of an offender. Each has distinctive colours and patterns.

Pollen

Pollen grain from plants at the crime scene can transfer to clothing. Washing may not remove them from the seams of the garment. The main features that distinguish one type of pollen from another are size, shape, colour and surface texture. The outer walls have pores and furrows. They can be meshed, granular, spined, or smooth.

> **e** A suspect has grass pollen on his trousers, identical to those in Figure 4. What, if anything, does this prove?

Dust and fibres

Dust contains shed skin cells, dust mites, hairs and fabric threads, soil and earth, pollen grains, fungal spores and particles of food. Three or more hairs fall unnoticed from your head every hour.

A combination of different multicoloured fibres, from a carpet, is conclusive evidence to match a suspect to a crime. The tell-tale fibres that garments shed are also distinctive. Wool has a pattern of surface scales. Silk and most synthetic fibres are smooth.

Figure 4 SEM pollen grain images from a crime scene: 1 dahlia, 2 grass, 3 chickweed, 4 poppy

Summary questions

1 Which clothing fibre, a, b or c, is the best match to the control sample? Explain your answer.

 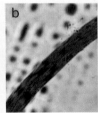  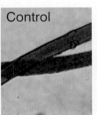

a   b   c   Control

2 Suppose you collect paint flakes at a crime scene and find some on a suspect. They do not seem to fit together, yet layers observed under a microscope are identical. What does this prove?

3 How does the appearance of pollen grains differ from that of fibres under a microscope?

Key points

- Forensic scientists use comparison, polarising and electron microscopes to improve contrast and resolution. This provides more precise and reliable evidence than simple light microscopes.

- The size, surface pattern and colour of pollen grains and layers of paint have distinctive features. Fibres have different colours, patterns or textures.

5.10 Modern analytical instruments

Learning objectives

- How has computer control changed scientific analysis?

- How do we interpret results obtained from analytical instruments? [H]

⊂⊃ links

For more information on ions look back at 5.2 Distinguishing different chemicals. For more information on chromatography look back at 5.8 Chromatography.

AQA Examiner's tip

In the exam you do not need to know details of the instruments described here. They are included for your interest. In the exam you will be expected to interpret traces to identify a match between an unknown and known substance.

Figure 1 Forensic drug detection apparatus. A gas chromatography machine (left) connected to a mass spectrometer (right).

Analytical scientists

Scientists of a past generation could only dream of working with the analytical techniques that are available to scientists nowadays. Thanks to advances in technology, reliable results are possible even with tiny quantities of materials.

Analytical chemists test air, water, industrial waste, drugs and food to make sure they are safe. They are employed by companies, including those producing food, pharmaceuticals, petrochemicals, agrochemicals, and other chemical and polymer manufacturers. They also work in hospital and public health laboratories, government and environmental agencies. They use high-tech equipment and sophisticated techniques to analyse substances and complex mixtures, often working at the current limit of technology.

The Forensic Science Service's toxicology team tests body fluids for poisons and illegal drugs. They deal with cases such as *driving under the influence of drugs* and *the use of drugs to aid sexual assault*. No longer do forensic scientists need weeks to analyse materials. Fast computers allow them to match complex data quickly – data from fibres, fingerprints, DNA and organic chemicals.

a How does technology help scientists to analyse chemicals more quickly nowadays?

Instruments for analysing drugs
Gas–liquid chromatography

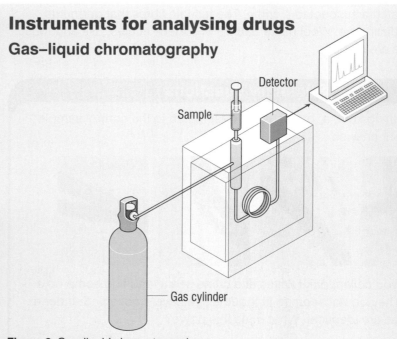

Figure 2 Gas–liquid chromatography

In gas–liquid chromatography a non-reactive gas, like nitrogen, flows through a narrow tube. A technician injects a vaporised drug sample into this flow. Chemicals in the sample move at different speeds to the detector. They appear as peaks on the computer display. A peak at the 'retention time' matching that of a known drug identifies the unknown substance.

b Why must tests be carried out on known drugs first?

Mass spectrometry

Scientists often separate a sample using gas chromatography before embarking on further analysis with a mass spectrometer. The two instruments are often linked together.

By heating the gas in a mass spectrometer the particles ionise. The ions accelerate into a magnetic field – lighter ions deflecting more easily. Scientists identify the composition of the sample from the range of ions detected.

c How does Figure 4 show that the athlete took an illegal steroid drug to enhance his performance?

Infrared spectrometry

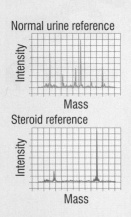

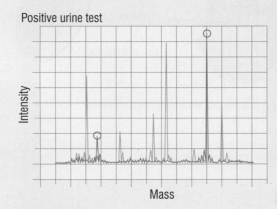

Figure 4 Mass spectrometry data from an athlete

After absorbing infrared radiation, samples produce distinct spectra or traces. The analyst uses a computer database of known traces to match against the traces from an unknown sample. Apart from identifying drugs, this method helps to identify plastics and paint samples.

d Look at Figure 5. Why can the forensic scientist say that the unknown sample is heroin?

e Why do modern instruments provide more precise and reliable evidence than that obtained from simple laboratory experiments?

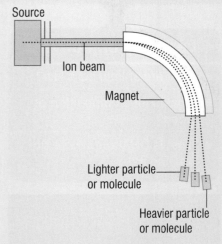

Figure 3 Mass spectrometer

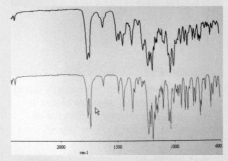

Figure 5 Infrared (IR) spectra of the illegal drug heroin (blue) compared with an unknown sample (black)

Summary questions

1 Explain the sentence: 'In gas–liquid chromatography, peaks at retention times matching those of a known drug indicate a positive result.' **[H]**

2 In mass spectrometry, what could you conclude if the mass of each of the ions in two samples were identical? **[H]**

3 Look at the infrared spectra (traces) in Figure 5. Comment on the match between the spectrum of the unknown substance and that of heroin. **[H]**

Key points

- Computers speed up forensic searches and make high-tech analysis of samples more precise and reliable.

- Analytical scientists use gas–liquid chromatography, mass spectrometry and infrared spectroscopy to identify unknown samples. An unknown substance can be identified by matching the distinctive features of its trace with those of known substances on a computer database.

5.11 | Blood and DNA

Learning objectives

● What are animal cells like?

● What is the composition of blood?

● What is DNA and where do you find it?

Figure 2 Revealing hidden blood – using a UV lamp to look for signs of blood. It is impossible for anyone to remove all traces of blood from a crime scene.

??? Did you know ... ?

The distribution of blood types varies from country to country because people have different ethnic origins. The table compares the percentage of the population in the UK and India with different blood types:

Country	A	B	AB	O
UK	42%	10%	4%	43%
India	23%	32%	7%	39%

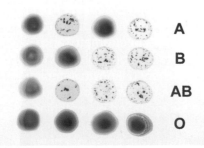

Figure 3 Blood group testing. The sample of blood sticks together if there is a positive test.

Animal cells

All cells are complex, allowing life to survive and reproduce. Animal cells have a nucleus, cytoplasm and cell membrane. Inside the nucleus you find **DNA**, i.e. **deoxyribonucleic acid**, which stores genetic information.

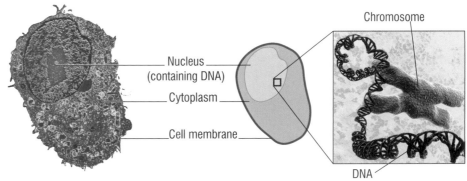

Figure 1 An animal cell with simplified drawing

a Where in a cell is the DNA?

Blood

You have over five litres of blood circulating round your body!

Blood contains a straw-coloured liquid called **plasma**. Plasma carries the red blood cells, white blood cells, platelets and various dissolved chemicals, such as nutrients, like glucose, round the body:

● **Red blood cells** don't have a nucleus. They contain **haemoglobin** to carry oxygen.

● **White blood cells** defend us against disease. They produce **antibodies** to fight bacteria.

● **Platelets** help the blood to clot at a wound.

Most people's blood belongs to one of the four main blood groups: **A**, **B**, **AB** and **O** (see Figure 3).

b What is the purpose of red blood cells?

c What can you say about the parents of a child of group AB?

Scientists test for blood groups by adding different antibodies to blood samples and noting if the blood clumps together. They can also identify whether or not a blood sample is human. Since it is quick and cheap, forensic scientists may carry out a blood test to eliminate someone from an enquiry. If their suspect and the blood samples from the crime scene do not match, they avoid the expense of a DNA test.

DNA

Coiled inside the nucleus of all the cells of your body is your DNA. Apart from identical twins, everyone's DNA is different. These metre-long DNA molecules consist of two strands wound around each other in a double helix. They contain our **genes** – the genetic traits we inherit from our parents. In fact we get half of our DNA from each parent. DNA testing can show whether or not people are related.

d Except for identical twins, what can we say about the DNA of every individual?

Practical

Extracting DNA from kiwi fruit

- Peel and mash up a kiwi fruit thoroughly.
- Mix $100\,cm^3$ distilled water, 2 g of salt, $5\,cm^3$ of washing up liquid avoiding bubbles. Stir in the kiwi mash.
- Heat in a water bath at 60 °C for 15 minutes then cool in ice water.
- Sieve the mixture into a beaker to remove lumps.
- Pour ice cold ethanol down the side of the beaker so it floats on the mixture.
- Pull the white stringy DNA from the layer with a splint.

A database is an electronic filing system, used to record and search for information. The **National DNA Database**, used in forensic science, stores millions of 'criminal justice' DNA profiles.

Summary questions

1 Describe the composition of **a** animal cells **b** blood.

2 The computer printouts of the DNA profiles A and B, shown in Figure 4, are a good match. What conclusion can you make?

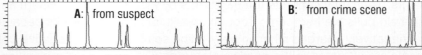

Figure 4 Computer DNA profiles – printouts A and B

3 By checking for similarities in the DNA profile patterns you can show whether or not people are related.

Look at the DNA profiles, C–E in Figure 5. What conclusion can you make?

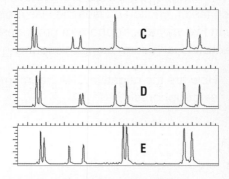

Figure 5 Computer DNA profiles – printouts C, D and E

Did you know ... ?

Family DNA link is a crime breakthrough

Using a 'familial DNA search', police successfully prosecuted a man who admitted killing a 53-year-old lorry driver, by throwing a brick through the windscreen of his cab.

The accused man and a friend had been drinking. Making their way home, they crossed a footbridge over the M3, where they hurled bricks at traffic. One smashed through the windscreen of a lorry. The brick hit the lorry driver in the chest, causing heart failure.

Scientific examination of the brick provided DNA evidence. A DNA profile was run against the National DNA Database, but no match was found. Yet 25 people nearly matched the profile. One of these lived nearby and he was a relative of the accused man. The accused man's DNA sample matched the one from the crime scene. He admitted manslaughter and was sentenced to six years in prison.

Key points

- Animal cells have a nucleus, cytoplasm and cell membrane.
- Blood contains red blood cells, white blood cells, platelets and plasma. The four main blood groups are A, B, AB and O.
- DNA is found in cell nuclei. You inherit DNA from your parents. DNA is unique to an individual (except identical twins).

5.12 DNA profiling 🔊 *k*

DNA

Apart from identical twins, no two people have the same DNA. You can extract samples of DNA from blood, saliva, semen and body cells (even hair follicles and dandruff). Forensic scientists analyse saliva from places such as cigarette ends and chewing gum. After sexual assaults they collect semen from vaginal swabs, underwear and beds, and body cells from under fingernails.

a What are the advantages and disadvantages of a DNA test compared to a blood test?

Electrophoresis

Scientists need only tiny samples of body cells to obtain DNA. From these samples scientists can identify their distinct DNA profile by a process called **electrophoresis**. You can think of electrophoresis as electrically forced chromatography.

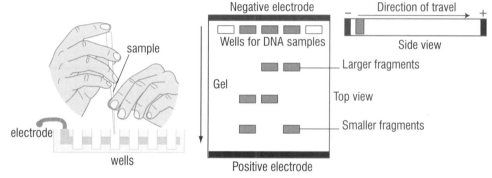

Figure 2 Preparing electrophoresis gel for DNA sequencing. Notice the electrical contacts and the rate of travel of the fragments.

To obtain a DNA profile, scientists use an enzyme to cut up the DNA into fragments. Then they separate the fragments by electrophoresis, where an electric field pulls the fragments through an alkaline gel. This works because DNA is negatively charged when in an alkaline solution, so the negative DNA fragments are attracted to the positive electrode. Different-sized molecules move through the gel at different rates, with the smaller molecules moving much faster. Friction causes the less mobile, larger fragments to stick to the gel. In this way electrophoresis produces a pattern of DNA bands unique to one individual human being (discounting identical twins).

b How are the negative DNA fragments forced through the gel in electrophoresis?

Learning objectives

- What is electrophoresis and how is it used?
- How can DNA fragments be identified using electrophoresis? [H]
- What are the ethical implications of storing DNA profiles?

Figure 1 DNA swab and forensic evidence form. You use a sterile cotton bud to collect cells from the lining of the mouth.

Activity

Different people hold different views about the ethical implications of storing DNA profiles:

'The innocent have nothing to fear from a DNA database of the whole population.'

'A UK-wide database erodes my civil liberties.'

- Discuss the issues.

Higher

Once the DNA fragments have separated, the pattern of bands on the gel is transferred by computer. Analysis software can establish matches and partial matches to other profiles held on its database and also produce graphical images for comparison.

Electrophoresis has uses other than in forensics. Analytical scientists use electrophoresis to diagnose genetic diseases and to test for a child's true parents.

Analysis of DNA profiles

Medical researchers and criminal investigators appreciate having computer printouts of graphs as confirmation of a match. In forensic investigations DNA profiles are more useful than blood tests as many people share the same blood type, but DNA is unique.

c Describe the features shown in Figure 3 that show a match.

Figure 3 DNA printouts, showing a match between a sample found at the crime scene and one taken from the suspect.

Summary questions

1 Give two examples of a very small body sample, from which scientists can identify an individual by electrophoresis.

2 Look at Figure 5. Which way does the negatively charged DNA fragment move? [H]

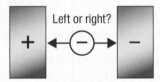

Left or right?

Figure 5 Electric field between plates

3 Figure 6 shows one type of electrophoresis equipment. The gel exerts a frictional resistance on the DNA fragments.

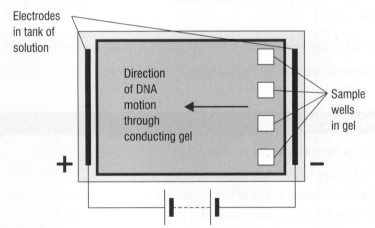

Electrodes in tank of solution

Direction of DNA motion through conducting gel

Sample wells in gel

Figure 6 Electrophoresis

a Which size fragments move more slowly through the gel?
b Why do the different size fragments move at different speeds?
c How is this similar to chromatography? [H]

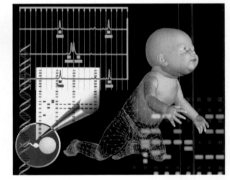

Figure 4 Paternity test. Any of the baby's DNA bands, or genetic markers, which do not match those of its mother must come from the father.

Key points

- Electrophoresis is like electrically forced chromatography. It separates the DNA fragments to produce a DNA profile. Everyone's DNA (like a fingerprint) is unique.

- In electrophoresis, negative DNA fragments in the alkaline gel are attracted to the positive electrode, with smaller fragments moving faster. [H]

- While some oppose our personal information being stored on national databases, others argue that it is only criminals who have something to fear.

5.13 Glass – the invisible evidence!

Learning objectives

- How does a glass surface refract light?

- How do scientists measure the refractive index of a glass fragment?

- How can trace evidence, like minute fragments of glass or plastic, be matched to a suspect?

Figure 1 Using forceps to collect a 'control glass sample' from a broken window. Will this sample match with fragments on the clothes or in the hair of a suspect?

Edmond Locard was known as the Sherlock Holmes of France. According to Edmond Locard, no criminal is invisible. 'Every contact leaves a trace.' The guilty always take identifying evidence away with them. They always leave traces of evidence at a crime scene – hairs, paint flecks, pollen, soil, fibres, and tiny flakes of glass or plastic.

We find tiny slivers of glass on suspects:

- vandalising or breaking and entering through windows
- involved in assault with bottles
- from their vehicle headlights in hit-and-run road traffic incidents.

Refraction

It is sometimes possible to match larger glass fragments together like a jigsaw. But how can we link a suspect to the crime scene with just a few slivers of glass or plastic? The **refractive index** (or the measure of the change of direction of a light ray) can help us out. Different types of glass have different densities, and so refract, or change the direction of the light, by different degrees. When light enters a more dense material it refracts towards the normal. The opposite is true if it goes into a less dense material.

We can find the refractive index of glass by measuring the angle of incidence (i) and the angle of refraction (r) (see Figure 2). Scientists calculate refractive index using the formula:

$$\text{Refractive index} = \frac{\sin i}{\sin r}$$

If light enters a transparent plastic at an angle of incidence of 40°, and the angle of refraction is 30°, the refractive index = sin 40° ÷ sin 30° = 0.643 ÷ 0.500 = 1.29.

a Which way does light refract if it emerges from a more dense material?

b What equation do we use to calculate refractive index?

Practical

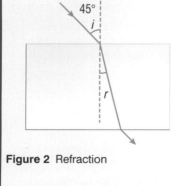

Line called the normal

45°

i

r

Figure 2 Refraction

Which refracts more, glass or Perspex?

1 Draw round a glass block on a sheet of white paper.

2 Use a protractor to mark a line at 45° as shown.

3 Direct a ray from a ray box along this line, then mark the ray that emerges from the glass.

4 Remove the glass block and complete the diagram as shown.

5 Measure the angle of refraction r.

6 Calculate the refractive index using the formula:

$$\text{Refractive index} = \frac{\sin i}{\sin r}$$

7 Repeat using the Perspex block.

- Which material has the greater refractive index?

We need a different method for measuring the refractive index of glass fragments – accurate results are just not possible with such small objects using sin *i* ÷ sin *r*.

Strangely, forensic scientists make glass fragments invisible to detect their refractive index!

Oil immersion method

Look at Figure 3. This is a photograph of a liquid droplet. Suppose the glass of the pipette and the liquid have the same refractive index. Then the light would not change direction at the glass–liquid surface. So you could not tell where the glass started and the liquid ended. As if by magic the surface is invisible!

To measure the refractive index of a tiny bit of glass, a forensic scientist immerses the fragment in an oil, under a microscope. The scientist then slowly heats and cools the oil. This changes the oil's density and refractive index. At just the right temperature the glass seems to disappear – when the refractive index of glass and oil match. A computer converts the temperature into a refractive index value.

Figure 3 A liquid droplet hanging on a pipette

> **c** Which method would you use to measure the refractive index of glass fragments from a crime scene and glass found on a suspect?
>
> **d** Explain why temperature affects the refractive index of oil.

Practical

Invisible glass

- Use a glass pipette. Fill three test tubes with glycerol.
- Warm two in hot water, taking one out earlier than the other. Your glycerol is now at three different temperatures.
- Immerse your pipette into each. Does the glass of the pipette all but disappear?

Figure 4 Glycerol at different temperatures

Practical

Refraction with a sugar cube

Suspend a sugar cube in water. Illuminate with light. As the sugar dissolves the denser sugar solution falls to the bottom. The solution has a different refractive index to the water. How can you tell?

Figure 5 Refraction with dense sugar solution

Summary questions

1 In Figure 6, which one of the arrows (A–E) shows the way the light will refract?

2 How does Figure 7 explain Locard's principle, 'Every contact leaves a trace.'?

3 Explain the oil immersion technique to measure the refractive index of a glass fragment.

4 In an experiment, the angle of incidence when a ray of light hits a glass surface is 60°, and the angle of refraction is 30°. What is the refractive index of the glass?

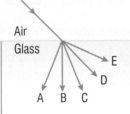

Figure 6

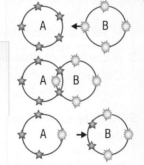

Figure 7

Key points

- Light refracts (changes direction) towards the normal when it enters glass or plastic.

- We can find the refractive index by using the formula **refractive index = sin *i* ÷ sin *r***

- Oil immersion method: a glass fragment disappears when put in oil at a certain temperature with the same refractive index.

5.14 Standard procedures for analysis

Learning objectives

- How do you remove solid matter to obtain a clear solution and test for the solubility of a compound in water?

- What analysis and detection standard procedures must you revise?

Quality control is important in food production. Analytical scientists may want to test the syrup in a fruit salad. The solubility of sugar in the syrup needs to be the same from one production run to another.

Practical

Filtering fruit salad and testing for the solubility of its sugar content

1 Filter some fruit salad, using a fluted filter paper, to obtain clear solution.

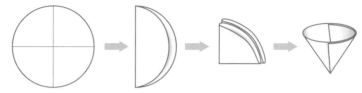

Figure 1 Separate small amounts of solid from a solution with a simple fold

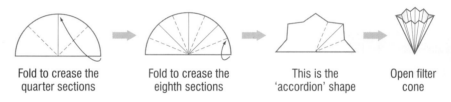

| Fold to crease the quarter sections | Fold to crease the eighth sections | This is the 'accordion' shape | Open filter cone |

Figure 2 Fluted filter paper lets liquid pass through quickly

2 Weigh an evaporating dish. Add 20 cm³ of clear solution to the evaporating dish. Place on a beaker of boiling water over a Bunsen flame. Stop heating when the solution goes tacky. What does it look like now?

Figure 3 Evaporating the solution

Maths skills

Check the following topic spreads for details of calculations involving:

- How to write a balanced symbol equation (5.6 Balanced equations).

- The amount of substance in a titration (5.7 Titrations).

- The formulae for retention factor (5.8 Chromatography) and refractive index (5.13 Glass – the invisible evidence!).

a Why don't you heat the solution directly while you evaporate it away?

3 Allow to cool. Reweigh the evaporating dish and its sugary contents. Calculate the mass of the sugary residue. Your answer is the mass of sugar in 20 cm³ (or 20 g) of fruit salad solution.

Solubility is the mass of a compound (in grams) that is dissolved per 100 g of water.

b You know the mass dissolved in 20 g, so what was the solubility of the sugar in the fruit salad (in units of grams of sugar per 100 g of water)?

Links to standard procedures

Use the following table to help you to revise the standard procedures.
Remember to carry out a risk assessment before carrying out any test.

Students need to know:	Topic	Further guidance
How to test for carbonate CO_3^{2-} ions	5.3	1 Wearing eye protection, add dilute hydrochloric acid. 2 Test for carbon dioxide using limewater, which turns cloudy.
How to test for alcohol (ethanol)	5.4	Wearing eye protection, add acidified potassium dichromate solution, which turns from orange to green in ethanol, when heated in a water bath. (No naked flames.)
How to carry out a flame test	5.5	1 Wearing eye protection, clean a nichrome wire loop in concentrated hydrochloric acid. 2 Dip the loop into the sample, place in the edge of a hot bunsen flame and record the colour.
How to measure pH	5.6	Wear eye protection and use universal indicator paper and compare to a colour chart (or use a pH meter).
How to carry out acid–base titrations to analyse the amount of substance in a sample	5.7	1 Wearing eye protection, pipette a measured volume of alkali into a conical flask and add indicator. 2 Fill a burette with acid and slowly add the acid to the alkali, swirling the flask. 3 Stop at the 'end point' (the indicator just changing colour) and record the volume of acid used. 4 Repeat until two titration volumes (V in cm^3) agree. 5 Calculate the moles of acid used to neutralise the alkali using ($V \div 1000$) × moles/dm^3. 6 Calculate the mass of alkali in sample using (moles used × relative formula mass of alkali).
How to analyse a chromatogram	5.8	1 Wearing eye protection, 'spot' the chromatography paper, leave for the solvent to rise and mark the solvent front before drying. (No naked flames.) 2 Measure and calculate the retention factor R_f to compare samples from the spots obtained. R_f = distance substance travels ÷ distance solvent travels
How to measure the refractive index of **a** a glass block and **b** a glass fragment	5.13	**a** 1 Shine a ray through the block. 2 Draw around the block and draw the rays. 3 Measure the angles of incidence and refraction. 4 Calculate the refractive index = sin i ÷ sin r **b** 1 Immerse the fragment in oil on a microscope slide. 2 Heat/cool gently until the fragment 'disappears'. 3 Use the temperature to calculate the refractive index.
How to obtain a clear solution	5.14	Wearing eye protection, filter the mixture through fluted filter paper to collect a clear solution.
How to test for the solubility of a compound in water	5.14	1 Wearing eye protection, evaporate 100 g of the solution over a water bath. 2 Calculate the mass of the residue. The solubility is the grams of compound dissolved per 100 g of water.

Summary questions

1 If 4.3 g of sodium chloride are dissolved in 24 g of water, what is the solubility of the solution (in grams of compound dissolved per 100 g of water)?

2 Why do scientists calculate the refractive index of a transparent material?

3 Why is the retention factor R_f used in chromatography?

Key points

● To remove solid matter, filter the mixture through fluted filter paper to collect a clear solution. To test for solubility evaporate 100 g of the solution over a water bath and calculate the mass of the residue. The solubility is the grams of compound dissolved per 100 g of water.

● Examples of analysis and detection standard procedures are in topic spreads 5.3 (testing for carbon dioxide using limewater), 5.4 (testing for ethanol using acidified potassium dichromate), 5.7 (how to carry out titrations to analyse the amount of substance in solution) and 5.13 (how to measure refractive index).

● Revise all the techniques listed in this topic spread including those referred to in the Maths skills box.

Summary questions

1 Copy and complete the sentences using the words below:

covalent   energy   giant   high
negative   organic   point   positive

The structure of ionic compounds is a .... lattice, which is held together by the attraction of .... and .... ions. Ionic compounds have a .... melting .... because you need a lot of .... to break the bonds between the ions. Substances obtained from living materials are .... compounds with .... bonding.

2 Copy and complete the sentences using the words below:

dioxide   ethanol   glucose   low
separate   strong   water   weak

CO_2 is carbon .... , H_2O is .... , C_2H_5OH is .... and $C_6H_{12}O_6$ is .... The covalent bonds between the atoms in a molecule are .... Covalent compounds have .... melting points because the forces between the molecules are .... and we only need a little energy to .... them.

3 Copy and complete the following word equation and write a balanced symbol equation for the reaction:

lead nitrate + sodium hydroxide $\longrightarrow$ ... + ... **[H]**

4 Copy and complete the sentences using the words below:

$CaCO_3$ $CuCl_2$ KCl Na_2SO_4 hydroxide
mass   mole   reacts   relative   titration

What is the formula of each of these?

potassium chloride .... , calcium carbonate .... , sodium sulfate .... , copper(II) chloride .... The .... formula .... of a substance, in grams, is known as one .... of that substance. The precipitate that forms when copper(II) sulfate .... with sodium hydroxide is copper .... You can analyse the acid content of rain water by ....

5 a Calculate the relative formula mass of potassium hydroxide (KOH).
(Relative atomic masses: H = 1, O = 16, K = 39)
b Suppose 0.01 moles of nitric acid (HNO_3) neutralise a solution of potassium hydroxide (KOH).
What mass of KOH does the solution contain?

6 Name an analytical instrument that is precise and produces reproducible data, and explain why these words apply to it. **[H]**

7 Copy and complete the sentences using the words below:

A   AB   membrane   mixture   nucleus
O   platelets   red   solvent   stationary
strongly   white   cells

In chromatography some colours in the .... don't move as far in the .... This is because they cling more .... to the .... substance they pass through. Blood contains .... and .... blood .... , plasma and .... The four main blood groups are .... , B, ... and .... Animal cells have a .... , cytoplasm and cell .... DNA is found in the nucleus.

8 What is the retention factor (R_f) for a substance that travels 6 cm when the solvent front travels 10 cm?

9 Explain chromatography using the words solute, solvent and medium. **[H]**

10 Explain how DNA fragments are separated by electrophoresis.

11 Copy and complete the sentences using the words below:

chromatography   distinctive   fibres   grains
layers   spectrometry   surface

Bullets, ... and pollen .... and paint .... can be distinguished by their .... features under a microscope, such as .... pattern. Forensic scientists use gas–liquid .... and mass and IR .... for drug testing in the crime lab.

12 Explain the oil immersion method to determine the refractive index of a glass or clear plastic fragment.

13 What is the refractive index of a clear plastic sample if angles of incidence and refraction correspond to 50° and 37°?

Key practical questions

1 Describe how to:
   a test for carbonate CO_3^{2-} ions
   b test for alcohol (ethanol)
   c measure pH
   d carry out acid–base titrations to analyse the amount of substance in a sample
   e analyse a chromatogram
   f measure the refractive index of a glass block
   g remove solid matter to obtain a clear solution
   h test for the solubility of a compound in water.

2 a How do you use precipitation reactions to detect the presence of:
   i the non-metallic ions chloride and sulfate (Cl^- and SO_4^{2-})
   ii the metal ions Ca^{2+}, Cu^{2+}, Fe^{2+}, Fe^{3+}, Pb^{2+}?
   b What are the products in the precipitation reaction between calcium nitrate and sodium hydroxide?

AQA Examination-style questions

1 Analytical scientists carry out qualitative analysis to identify the presence of chemicals in a sample of white crystals. The scientist carries out a flame test to identify the sample. He dips a wire loop in the sample which he then holds in the flame.

a Why would the scientist dip the wire loop in acid and then put it in the flame before dipping it in the sample of white crystals to be tested? *(1)*

b What would the scientist see to confirm the presence of sodium ions in the sample of white crystals when carrying out the flame test. *(1)*

c Sodium chloride was detected in the white crystals. Name the type of bond present in sodium chloride. *(1)*

2 Scientists have been asked to find out why fish have been dying in the local pond. The local people are concerned someone is dumping acid in the lake.

a The scientists used universal indicator paper to see if the lake water was acidic. What result would they get if the lake water was acidic? *(1)*

b The scientist carried out a titration to test how acidic the lake water was. They titrated 25 cm³ of the water against sodium hydroxide.

  i What piece of equipment would he use to accurately measure 25 cm³ of water from the lake? *(1)*

  ii Name the piece of equipment he would use to add the sodium hydroxide to the water? *(1)*

  iii What is the relative formula mass of sodium hydroxide (NaOH)? (A_r H=1, O=16, Na=23) *(1)*

c The equation for the reaction between nitric acid and sodium hydroxide is shown below.

$$HNO_3 + NaOH \rightarrow NaNO_3 + H_2O$$

Nitric acid has a relative formula mass of 63. Sodium nitrate has a relative formula mass of 85. What mass of sodium nitrate would be produced if 12.6 g of nitric acid reacts with excess sodium hydroxide? *(3)*

3 A bullet had been fired at the scene of a crime.

a Describe a test a forensic scientist could carry out on the bullet that would allow it to be matched to the gun that had fired it. *(3)*

b The forensic scientist examined a sample of the blood from the crime scene under a microscope. The diagram shows what he saw.

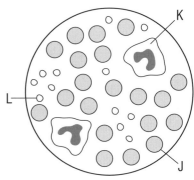

  i Name **J**, **K** and **L**. *(3)*

  ii Which part of a blood cell is needed for DNA profiling? *(1)*

  iii The forensic scientist also tested the blood to identify its blood group. Name the **four** main blood groups. *(2)*

AQA, 2008

4 Analytical scientists found some glass at the scene of a hit-and-run. The glass was examined to find its refractive index.

a The diagram shows a ray of light entering the piece of glass.

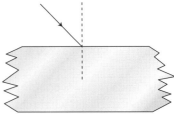

  i On the diagram, draw the path of the ray of light as it passes into and then out of the glass. *(2)*

  ii On the diagram, label the angle of incidence (*i*) and the angle of refraction (*r*). *(2)*

  iii Explain how the refractive index of a small fragment of glass is obtained? *(3)*

b A suspect was arrested and breathalysed. The breathalyser test detects a compound with the formula C_2H_5OH.

  i What is the name of this compound? *(1)*

  ii Name the type of bonding in this compound. *(1)*

  iii Describe how a forensic scientist might test for this compound in the laboratory. *(2)*

  iv Suggest one method why the modern breathalyser is more accurate than those first used. *(2)*

AQA, 2008

1 Environmental scientists use forensic techniques to investigate how human activities affect the environment. An environmental scientist collected a sample of river water near an industrial estate to see if there was any contamination. The environmental scientist filtered the sample and tested the filtrate and residue.

a The environmental scientist wants to test the filtrate from the river for its pH. Suggest why the scientist would use a pH meter rather than using universal indicator paper. *(2)*

b i The environmental scientist carried out a number of tests on the residue. Choose the correct words to complete the sentences below.

hydrochloric acid   limewater   cloudy   clear   oxygen   carbon dioxide hydrogen

_____ was added to the residue. The gas given off was collected. The gas was bubbled through _____ which becomes _____ . This shows that _____ has been given off. Therefore the residue is a carbonate. *(4)*

ii Some solution from **b i** is added to sodium hydroxide and a blue precipitate is observed. State the name of the substance. *(1)*

2 An analytical chemist, working for a bank, has been asked to examine the ink found on a forged signature on a cheque. He used paper chromatography to compare the ink on the cheque with ink from 3 brands of pen.

a Choose the correct answer to complete the sentence.

compound   molecule   mixture   element *(1)*

Chromatography is a method used to identify chemicals in a ………

b *In this question you will be assessed on using good English, organising information clearly and using specialist terms where appropriate.*

Describe the method the scientist would use to carry out the chromatography and explain how he would know which pen had been used to forge the signature on the cheque. *(6)*

3 The labels have come off several containers of chemicals in a research lab. An analytical scientist has been asked to analyse the substances and label the containers correctly.

a The scientist finds two jars containing crystals. He decides to carry out a flame test on each substance to find what metal ions are present. He dips a nichrome wire loop in acid before dipping it in the powder. He then holds the wire loop in a flame.

His results are shown in the table below. Complete the table by filling in the metal ion present in each substance.

Substance	Flame colour	Metal ion present
First crystal sample	Blue/green	
Second crystal sample	Bright yellow	

(2)

b He adds hydrochloric acid to the first crystal followed by barium chloride solution. A white precipitate is observed. Name the anion present. *(1)*

c He thinks the second crystal may be a chloride. Describe a test he can do to prove that he is correct. *(3)*

d Both the crystals are ionic compounds. Describe the structure of an ionic compound. *(2)*

e Give the name of each crystal sample. *(2)*

AQA Examiner's tip

This question may be about a situation you are not familiar with or have not studied. Sometimes exam questions can be based on situations that are new to you. However, they will always be based on knowledge you should have studied. By basing the questions in a new situation it shows the examiner how much you have understood the science you have studied and whether or not you can use that knowledge.

AQA Examiner's tip

Make sure when you are answering the extended questions that you use the appropriate scientific terms. It is important for question 2b you both describe the method and explain how to use the results. If you a miss a part of the question you will get low marks.

4 Police have found a child who they think was kidnapped 5 years ago. The child doesn't remember his real parents, and the police ask forensic scientists to help them find out who he really is.

a The forensic scientists ask the parents of the kidnapped child to provide a blood sample. A DNA profile is created from each blood sample.

Using the DNA profiles shown in the diagram, decide if the man and woman tested are the parents of the child, explaining why.

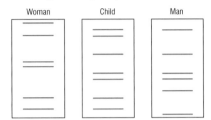

Woman   Child   Man

(2)

b Name the technique used to create the DNA profiles. (1)

c *In this question you will be assessed on using good English, organising information clearly and using specialist terms where appropriate.*

Describe and explain how the DNA profile is created, using the apparatus shown in the diagram. **[H]** (6)

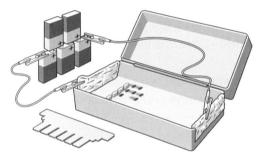

5 An analytical scientist was testing the boiling points of some compounds. The scientist found that all the ionic compounds had high boiling points and all the covalent compounds had low boiling points.

a i Explain why the ionic compounds have high boiling points. (2)

ii Explain why the covalent compounds have low boiling points. (1)

b The table shows the names and formulae of some covalent molecules. Copy and complete the table below.

Name	Formula
carbon dioxide	
water	
ethanol	
	$C_6H_{12}O_6$

(4)

6.1

Investigating the work of scientists and how they use science ⓚ

Learning objectives

● What aspects of Assignment 1 are critical?

● How should I set out my report?

∞ links

For information and examples of healthcare, materials, food and analytical scientists look back to the Making connections spreads for for Chapters 2, 3, 4 and 4.1 Introduction to food science.

For more information and help with research see 6.2 Maximising your marks.

For information on following standard procedures look back to 1.4 Following standard procedures and 1.5 Carrying out a standard procedure.

For information on standard procedures in healthcare, materials, food and analytical science look back to 2.12 Standard procedures for maintaining health and fitness, 3.8 Standard procedures for testing materials and 4.16 Standard procedures used in food science.

Introduction

Assignment 1 starts with research into the work of a healthcare, materials, food or analytical scientist. You then do a practical investigation based on a standard procedure that the scientist uses. This assignment is worth 40 marks and takes about 20 hours to complete. This time includes your preparation and research. Your work must always be your own and never copied. You cannot take your work home, not even on a memory stick.

Checklist for your assignment report

Set your work out under headings:

Assignment 1 title –This must be based on which option you choose from the four set by AQA.

Research report – worth 23/40 of the marks available

● **The scientist's employer** – Explain why the organisation exists. Describe the work done *and* how the organisation uses this to benefit society.

● **The scientist's work** – Describe their work. Include examples of activities from Chapters 2–5.

● **Qualifications and skills** – Describe the scientist's qualifications. Explain how their scientific knowledge *and* practical skills would help them with this investigation.

● **Sources of information** – State why it is possible that conclusions from this investigation may be unsound *and* suggest better ways to collect the data. Complete a *bibliography* as you research and write.

Investigation report – worth 17/40 of the marks available

● **Hypothesis** – Write a hypothesis for your investigation. Do not forget to give reasons for any assumptions made in the proposal.

● **Standard procedure** – Record results carefully in tables and graphs. Give reasons for repeating any measurements.

● **Conclusions** – Do necessary calculations to analyse your data. Explain patterns in the data *and* write your scientific conclusions.

Figure 1 Student research

AQA Examiner's tip

Explain how skills assist your scientist in their work.

General skills needed	Practical skills needed	Qualities needed
Interpersonal	Able to hypothesise	Patience
Communication	Hand–eye coordination	Reliability
Numeracy	Observational	Initiative
Organisational	Measuring	Flexibility
ICT	Awareness of anomalies	Accuracy
Imagination	Health and safety	Caring attitude

Marking grid for Assignment 1

The following is a student friendly version of the AQA marking criteria (40 marks in total):

Level 1	Level 2	Level 3
1. Research 1A. Information on the organisation		
You **state** the purpose of the organisation where the scientist works, **and** the purpose of the investigation. Your work is poorly **organised** but contains **some valid points**. (1–2 marks)	You **describe** the purpose of the organisation where the scientist works, **and** the purpose of the investigation. Your work has some **structure** and contains **some valid evidence**. (3–4 marks)	You **explain** the purpose of the organisation where the scientist works, **and** the purpose of the investigation, describing their benefits to society. Your work is **logically structured** with a **range of valid evidence**. (5–6 marks)
1. Research 1B. Information on the work of the scientist		
You **state** the work of the scientist, **adding one** scientific fact from the specification. (1–2 marks)	You **describe** the work of the scientist **adding some** scientific facts from the specification. (3–4 marks)	You **describe in detail** the work of the scientist adding **various scientific** facts from the specification. (5–6 marks)
1. Research 1C. Qualifications of/skills used by the scientist		
You **state** the qualifications of the scientist **and one** practical skill needed in the investigation. (1–2 marks)	You **describe** the qualifications of the scientist and **some practical** skills needed in the investigation. (3–4 marks)	You **describe** the qualifications of the scientist **and explain** how practical skills **and** scientific knowledge are used in the investigation. (5–6 marks)
1. Research 1D. Sources of information		
You use a **few sources** of information, some of which are provided for you. (1 mark)	You **record sources** of information you have selected. You **state the limitations** of the data and conclusions. (2–3 marks)	You write a **bibliography with many relevant sources** of information. You also **state alternative strategies** to improve the investigation data. (4–5 marks)
2. Making a hypothesis		
You state an **unscientific hypothesis**. (1 mark)	You state a **relevant hypothesis**. (2 marks)	You give **scientific reasons** for your hypothesis. (3 marks)
3. Following standard procedures and collecting data		
You try to follow the standard procedure. You **state observations** and **record results** in tables. (1–2 marks)	You follow the standard procedure. You **state observations** and **record accurate results** in tables **and** graphs, **repeating** results as needed. (3–5 marks)	You follow the standard procedure **by yourself**. You **state observations** and **record accurate, precise results** in tables **and** graphs, giving **reasons for repeating** results. (6–8 marks)
4. Analysing data/evidence and drawing conclusions		
You **try** to state patterns, do calculations and explain conclusions, but your work is disorganised. (1–2 marks)	You state **patterns**, do **calculations** and **explain conclusions** that **relate to the evidence**. (3–4 marks)	You **explain patterns**, do **complex calculations** and **explain conclusions** using **scientific understanding**, that **relate to the evidence**. (5–6 marks)

a What topic area interests you most – healthcare, materials, food or analytical science?

b Which standard procedures in the Assignment 1 options will you be able to follow successfully?

c Which scientist have you decided to research? Explain whether this scientist classifies, obtains or makes things, or tackles specific problems, or monitors and controls changes.

Key points

- Quality research and understanding the standard procedure are vital in Assignment 1.
- Set out your report clearly. Use the checklist headings.

6.2 Maximising your marks

Figure 1 Gathering information

Finding and saving information

To be able to write your report you need information. Thanks to the internet there is a huge amount of information available, but you should not rely completely on websites. Go to the library to use books and magazines. If you have the opportunity, write to or interview a scientist working in the field covered by your investigation.

Save any useful information obtained during your research, noting its bibliography reference. Organise your work under section headings as you go along.

a Why is saving useful sources of information as important as finding them in the first place?

Figure 2 Interviewing a scientist

Figure 3 Surfing the internet

Websites

If you use a lot of downloaded pages in your report, even if key sentences are highlighted, you will get no credit. The report must show your ideas and your own words, not someone else's. Websites are a good starting point for gathering information but you need to understand and explain your research.

Activity

Internet research

A good place to start your research is to explore the government's 'Next Step' careers advice website for job profiles. Under 'science and research' you will find information about healthcare, materials, food and analytical scientists.

Important: Do not limit your research to this website. Find a range of useful information.

You may already be familiar with using internet search engines, and how to use key words to refine your search. When performing a search, ask yourself the following questions:

- Which organisation constructed this website? So is it likely to be useful?
- Is the information reliable and up to date?
- Does the title of the page tell me what it is about?
- Do headings and illustrations help my understanding?
- Does the page give links to other useful sites?

When looking for information, learn to scan or skim-read the page, rather than trying to read every word. However, look more closely at the headings and information in figures and tables. Ignore any page that is difficult to understand or difficult to find information on. Ask yourself, 'Would I recommend this site to a friend?'

b What advice on searching the internet is most useful and why?

You've heard it said 'A picture is worth a thousand words', so don't forget about including suitable charts and images in your report. But don't be a copy-cat! Make all work your own.

c See the Activity box on internet research. How does the 'Next Step' careers advice website emphasise the importance of skills?

Your report

- Do not leave writing your report until you've finished all your research and investigating. Have a title page and complete the Research report and Investigation report sections as you go along.
- When writing, put your work onto the correct pages and add sub-headings as necessary.
- At the end of each research paragraph, write a number in brackets. For example: (4). Then match that number with its reference on the bibliography page.
- If you type up your report on a computer in school, do a spelling and grammar check before handing your work in.

Sections	Main headings	Sub-headings
Research report	The scientist's employer	• Why the organisation exists • The investigation • Benefit to society
	The scientist's work	
	Qualifications and skills	• Qualifications • Knowledge and skills
	Sources of information	• Investigation concerns • Bibliography
Investigation report (which must be linked to your research section)	Hypothesis	
	Standard procedure	• Results • Explaining repeats • Graphs
	Conclusions	• Calculations • Patterns • Conclusions

AQA Examiner's tip

- Be concise. Only use relevant information from your research.
- Make sure you can follow the standard procedure before you start.
- Keep your work structured. Use section headings.

?? Did you know …?

Scientists write bibliographies in a set order:

For books and magazines:
Author (date). Title. Publisher
For example, Whitaker et al. (Sept 2007) *Forensic entomology*. Catalyst p.1

From the internet: Title. URL
For example, Forensic Science Careers www.kent.ac.uk/careers/forensicsci.htm

AQA Examiner's tip

How long should you make your report?
It doesn't have to be long, you should:

- Stick to the main headings and sub-headings.
- Aim to pick up as many marks as possible.
- Refer to the checklist and marking criteria in 6.1 Investigating the work of scientists and how they use science.

Key points

- You can get information for your report from books, magazines, interviews and websites.
- You must use your own ideas and words in your report. Don't just repeat what you found in your research.

7.1

How scientists use evidence to solve problems ⓚ

Learning objectives

- How important is the Assignment 2 investigation?
- How should I set out my report?

∞ links

For information on how to achieve good marks, see 7.2 Maximising your marks.
For more information on writing risk assessments look back to 1.2 Risk assessments.
For information on following standard procedures look back to 1.4 Following standard procedures and 1.5 Carrying out a standard procedure.
For information on standard procedures in healthcare, materials, food and analytical science look back to 2.12 Standard procedures for maintaining health and fitness, 3.8 Standard procedures for testing materials and 4.16 Standard procedures used in food science.

Introducing the investigation

This practical investigation is worth 50 marks. The investigation is set by AQA and it gives you the opportunity to solve a scientific problem. You will learn about:

- the techniques these scientists use in their work
- the purpose of each technique and how it works
- the importance of assessing hazards and risks, and working safely and accurately
- collecting data from primary and secondary sources
- interpreting results and drawing conclusions
- evaluating methods of data collection
- presenting evidence and considering its reproducibility.

> **a** Primary sources provide data you obtain from your own investigations. Scientists collect secondary data from other scientists' investigations. Where could you find secondary data?

Setting the scene

The instructions from AQA set the scene for the basis of your investigation. This could be a brief (or request) from a client to you as a healthcare, materials, food or analytical scientist. The investigation has three phases: planning and risk assessment, practical work and data collection, and report writing.

The following is an example taken from forensic science.

> **The case of the vindictive vandal**
>
> Local farmer Terry Rout has had his garage broken into and his lovingly restored vintage car vandalised. The vandal also appears to have poisoned his trout lake.
>
> The crime scene investigator comments; 'It looks like the vandal smashed a hammer into the bonnet and the windscreen. There are no fingerprints or footprints. The hammer was clean and the floor of the garage was swept with a broom. There are no prints on that either. I recovered some control paint fragments from the bonnet, glass fragments from the windscreen and a sample of the thick dust from the floor. I also collected a sample of water from the lake.'

In this investigation you, acting as the forensic scientist, must examine the evidence collected at the crime scene and compare this to evidence collected from three suspects.

You are required to:
- Use a microscope to examine paint and dust samples found at the crime scene and from the clothes of the suspects.
- Test for ions present in the lake and from traces on the suspects' trousers.

- Check for any matching refractive indices from the windscreen and any slivers of glass found on the suspects.
- Write your report, with conclusions, based on the evidence obtained.

b Why is it important to follow standard procedures?

Checklist for your assignment report

Set your work out under headings, using ICT facilities if need be:

Title

Give your assignment a title based on your investigation scenario.

Planning

- **The purpose of the investigation**
- **Planning** – a series of well-ordered steps you are going to take for each task.
- **The equipment needed** – for each task.

Assessing and managing risk

- **Risk assessments** – hazards, risks, control measures, emergency action tables for each task.

Collecting data/evidence

- **Results** – for each task, including results tables (with averages) and photographs (where necessary), with comments about inconsistent (or anomalous) results.
- **Graphs** – labelled, like their corresponding table, where necessary.

Processing primary and secondary data/evidence

- **Calculations** – using averages (but rejecting anomalous results) and formulae.
- **Patterns** – explain the patterns in the data clearly.

Analysing primary and secondary data/evidence

- **Conclusions** – based on both the primary and secondary scientific evidence.
- **Validity** – say if your conclusions are well-grounded or not and how to improve your methods.

Evaluating the practical activity

- **Strengths and weaknesses** – effectiveness of methods used for each task and reliability (repeatability and reproducibility) of the evidence.
- **Improvements** – explain ways to improve your methods. (State each improvement and explain the reason, linking the sentence with the word 'because'.)

Workplace context

- **Research** – to explain the link between its application in the workplace and the investigation.
- **Summary** – your final conclusion and analysis, explaining how a scientist would use your results.

Figure 1 Student investigation

AQA Examiner's tip

Keep your work tidy and in order. Use the checklist to organise your work under headings.

Key points

- Assignment 2 is worth 50 GCSE marks. Along with Assignment 1, these tasks account for 60% of your GCSE marks.

- Set out your report clearly. Use the checklist headings.

7.2

Maximising your marks

● How can you obtain good marks in Assignment 2?

Figure 1 Feedback to students during assignment

∞ links

For a checklist giving headings and sub-headings for your report, look back to 7.1 How scientists use evidence to solve problems.

AQA *Examiner's tip*

Remember to:
● Write clearly in good English.
● Use the word 'because' in explanations, as this makes you explain your reason.
● Use bullet points for lists as they are an effective way to present information.

Dos and don'ts

Your teacher will introduce the investigation, suggest research sources, comment on the materials supplied and remind you about safe working practices.

You can do practical work in pairs or small groups, but all your work – your plan, risk assessment, results, etc. must always be your own and never copied.

You cannot take work home with you during the investigation, not even on a memory stick.

Achieving good marks

To achieve the best marks in Assignment 2, you should do the following:

1 Planning

State the purpose of the investigation and the tasks you will carry out. List the equipment needed for these tasks. Each task should have a detailed step-by-step plan, which includes the readings or observations needed for the task.

2 Assessing and managing risk

Describe the risks associated with each hazard and their control measures, if possible set out in risk assessment tables.

3 Collecting data/evidence

Record results and repeat readings (with averages) in suitable tables labelling tables and graphs correctly. Point out anomalous (inconsistent) results and explain why repeating certain results is necessary.

4 Processing primary and secondary data/evidence

Calculate averages (but reject anomalous results). Describe and explain the patterns in your data based on the evidence obtained. (Where possible use formulae to do appropriate calculations.)

5 Analysing primary and secondary data/evidence

Write conclusions based on the evidence and say how you could have improved your methods. Refer to secondary data, giving evidence of your investigation that supports your findings. Where your conclusions are not sound, consider other methods to improve the reproducibility of your data.

6 Evaluating the practical activity

Say how effective your methods were, stating both strengths and weaknesses. Explain, giving reasons, how you can improve your methods.

7 Workplace context

Describe the link between your investigation and the workplace, and how a scientist would use your results. Explain this link using information based on their research.

Marking grid for Assignment 2

The following is a student friendly version of the AQA marking criteria (50 marks in total):

Level 1	Level 2	Level 3
1. Planning		
You state the **purpose** of the investigation in a **plan** and write **some equipment** needed. Your work is **disorganised**. (1–2 marks)	You state the **purpose** of the investigation in a **plan** and **list the equipment** needed. Your work is **clearly structured**. (3–4 marks)	You state the **purpose** of the investigation and list **all the equipment** needed. Your **plan** is logically structured in **well ordered steps**. (5–6 marks)
2. Assessing and managing risk		
You refer to **health and safety** in a basic risk assessment. (1–2 marks)	You state **most hazards, risks and control measures**, based on common sense. (3–5 marks)	You state **the hazards, risks and control measures**, based on **scientific facts**. (6–8 marks)
3. Collecting data/evidence		
You state **basic observations** and record results in a **simple table**, with a chart or **graph**. (1–3 marks)	You state **accurate observations** and record **repeat** results in a **table** with 3 or more columns, with a chart or **graph**. (4–7 marks)	You state **accurate reliable observations** and record **repeat** results (with **consistent significant figures**) in a **logical table** of 3 or more columns, with a chart or **graph**. You identify **anomalous results**. (8–11 marks)
4. Processing primary and secondary data/evidence		
You make **mean calculations** and state **simple patterns**. (1–2 marks)	You make **calculations** to appropriate **significant figures** and **omit anomalous readings**. You **describe patterns** and **relationships**. (3–5 marks)	You make **complex calculations** to appropriate **significant figures** and **omit anomalous readings**. You **explain patterns** and **relationships**. (6–8 marks)
5. Analysing primary and secondary data/evidence		
You state **conclusions**, based **vaguely** on your evidence. You do not refer to secondary data. (1–2 marks)	You state **conclusions**, based directly on your evidence. You refer to **secondary data**. You state how to **improve the data** recorded. (3–4 marks)	You state **logical conclusions**, based directly on the **primary and secondary scientific evidence**. You state the limitations of your conclusions. (5–6 marks)
6. Evaluating the practical activity		
You write a **basic evaluation** and state **one improvement**. You have **poor spelling, punctuation and grammar**, and use few technical terms. (1–2 marks)	You describe the **success** of **methods** in your evaluation and **justify improvements**. You use some **technical terms**. (3–4 marks)	You **explain** the **strengths and weaknesses** of methods in your evaluation, and fully justify improvements. You use technical terms and have good spelling, punctuation and grammar. (5–6 marks)
7. Workplace context		
You state a simple **workplace application**. (1 mark)	You **describe** a workplace application and **state how** to use your results. Yet your opinion may need **more evidence**. (2–3 marks)	Based on **research** you **describe** a workplace application and **explain scientifically how** to use your results. (4–5 marks)

Key points

- To get good marks complete all the sections of your report, following the advice given on this page.

Glossary

A

Absorbs Soaks up. After food has been digested it is absorbed into the bloodstream.

Activation energy Energy required to start a chemical reaction.

Actual yield The mass of useful product gained from a chemical reaction.

Aerobic respiration The process by which food molecules are broken down using oxygen to release energy for the cells.

Agricultural scientist Scientist working in the agricultural sector, for example, one dealing with the selection, breeding and management of crops and animals.

Alloy A mixture of two or more elements, of which at least one is a metal.

Alveoli The tiny air sacs in the lungs which increase the surface area for gaseous exchange.

Anaerobic respiration The process of breaking down food without oxygen to release energy for the cells.

Antagonistic pairs Skeletal muscles that always work in pairs, like your biceps and triceps. When one contracts the other relaxes.

Antibody Protein made by the immune system to combat a disease.

Artery Blood vessel that carries blood away from the heart. (A vein takes blood to the heart.)

Aseptic Without the presence of micro-organisms.

Atom The smallest part of an element.

Atria The upper chambers of the heart. (One atrium, two atria.) (See also Ventricle. There are also two ventricles or lower chambers, the heart being a double pump.)

B

Baseline measurement Standard test to assess fitness of athletes.

Basic energy requirement (BER) For every kilogram of body mass you need 5.4 kJ (1.3 kilocalories) every hour.

Biodiversity The number and variety of different organisms found in a specified area

Biological control Using living organisms (predators of the pests) to control a pest population.

Bladder The organ that fills with urine from the kidneys, where it is stored before being released from the body.

Body mass index (BMI) To assess if people are under- or overweight. Your $BMI = \dfrac{weight}{height^2}$ (in kg/m^2).

Breathing rate The number of breaths per minute.

Brittle Cracks or fractures when a force causes stress.

BSI British Standards Institution that sets guidelines or standards that all companies who design and make things should follow for good practice.

C

Capillary Thin-walled blood vessel that exchanges substances with body cells as blood passes by.

Carbohydrate Compound containing carbon, hydrogen and oxygen, which is a major source of energy in animal diets.

Cartilage Smooth, slippery tissue covering the ends of bone that prevents bones from rubbing together.

CE The 'CE mark' is needed by many products in order for them to be bought and sold in the European Union. CE stands for conformité européenne, French for 'European conformity'.

CEN European Committee for Standardisation sets standards for Europe to make sure that products are safe to use and more reliable. European standards are adopted as British Standards.

Ceramic Inorganic non-metallic material such as pottery (made of clay) or glass.

Chromatography The process whereby small amounts of dissolved substances are separated by running a mobile solvent along a stationary substance, such as absorbent paper.

Comparison microscope Two microscopes used to analyse side-by-side specimens.

Composite Two or more materials mixed together, like glass-reinforced plastic for canoes.

Compression A force that pushes or squashes an object (rather than stretching it).

Compressive strength The ability to resist crushing.

Conducts Heat transferring in a substance due to motion of particles. (See also electrical and thermal conductors.)

Contrast Differences in the brightness and colour of an image that make detail clearer.

Controlled environment The environment in which an organism lives which is carefully controlled in order to maximise growth.

Core body temperature The internal temperature of the body.

Covalent compounds Chemical compounds formed through the sharing of electrons between atoms.

Cross-links Cross-links are bonds (or branches or bridges) that link one polymer chain to another, creating a thermosetting polymer (as opposed to a thermoplastic polymer, that doesn't have cross-links).

D

Defra The Department for Environment, Farming and Rural Affairs that monitors farming in the UK.

Density Mass per unit volume of a substance.

Deoxygenated blood Blood that is lacking in oxygen, on its way back to the lungs.

Diabetes A condition in which it becomes difficult or impossible for your body to control the levels of sugar in your blood.

Diaphragm The sheet of muscle which divides the thorax from the abdomen, used to aid ventilation of the lungs.

Diet diary A record of a person's food and drink intake over a period of time.

Dietician A scientist who advises people about the foods they should eat and those they should avoid.

DNA Deoxyribose nucleic acid, the material of inheritance.

Double pump A vessel containing two pumps that work together, like the atria and ventricles in the heart.

Dynamic equilibrium The state reached in a reversible reaction when there is a fixed ratio between the production of products and reactants.

E

Elastic limit The point beyond which a spring (or wire) is no longer elastic, and doesn't return to its original length.

Electrical and thermal conductors Electrical conductors, like metals, allow electrons to flow though them. Good electrical conductors are also good heat conductors, as the electrons help with passing energy from atom to atom.

Electrolyte A liquid containing dissolved ions.

Electron A tiny negatively charged particle that orbits the nucleus of an atom.

Electrophoresis A technique that uses an electrostatic force to separate large molecules-such as DNA fragments.

End point The designated point in a chemical reaction that defines the end of the reaction.

Endothermic A reaction that takes in energy from the surroundings.

EPS Expanded polystyrene.

Error Sometimes called an uncertainty.

Eutrophication The process by which excessive nutrients in water lead to very fast plant growth. When the plants die they are decomposed and this uses up a lot of oxygen so the water can no longer sustain animal life.

Evaporation Energetic liquid molecules turning to a gas below its boiling point.

Exothermic Releases heat energy.

F

Fermentation The reaction in which the enzymes in yeast turn glucose into ethanol and carbon dioxide.

Fertiliser A substance that is added to the soil to aid plant growth.

Fibre A food group which is not digested by the body, but is important to help remove waste products from the body.

Fit for purpose A product which meets a particular need or specification.

Flammable Capable of burning easily.

Flexible Not stiff

Food poisoning An infection caused by food containing harmful bacteria.

Food scientist A scientist who is involved in the production, manufacture or analysis of food.

Fungicide A chemical used to kill fungi.

G

Gene A short section of DNA carrying genetic information.

Gene pool The total genetic information present in a breeding population.

Genetic engineering/modification A technique for changing the genetic information of a cell.

Glucagon Hormone involved in the control of blood sugar levels.

Glucose-testing strip A dip-stick to determine the concentration of glucose in urine or blood.

Glycogen Carbohydrate store in animals, including the muscles, liver and brain of the human body.

H

Haemoglobin The red pigment which carries oxygen around the body.

Hard Difficult to dent or scratch.

Hazard A hazard is a potential danger (for example, an object, a property of a substance or an activity) that can cause harm.

Health and Safety Executive The organisation responsible for the regulation of risks to health and safety in the workplace.

Health Protection Agency An independent body that protects the health and well-being of the population.

Herbicide A chemical used to kill weeds or other plants with leaves.

Hooke's Law The extension of a spring is directly proportional to the force applied, provided its elastic limit is not exceeded.

Hydroponics Growing plants in water enriched by mineral ions rather than soil.

Hypertonic High in glucose, as in energy-based sports drinks.

Hypotonic Low in glucose, as in rehydrating sports drinks.

I

Insulin A hormone involved in the control of blood sugar levels.

Intensive farming A method of farming to minimise costs yet maximise production.

Intercostal muscle Muscle lying between the ribs, that pull the ribs upwards and outwards to aid breathing.

Ion A charged particle produced by the loss or gain of electrons.

Ionic compounds Chemical compounds formed through the donation and receiving of electrons.

Isotonic Having the same concentration of ions as another liquid, such as the blood.

J

Joint The place where bones connect, allowing motion.

K

Kidney An organ that filters the blood and removes urea, excess salts and water.

L

Lactic acid One product of anaerobic respiration. It builds up in muscles with exercise. Important in yoghurt and cheese making processes.

Level of water The human body functions best when the proportion of water is about 60 per cent.

Lever A system of applying a force (or effort) to another object (or load) by using a pivot.

Ligament Strong, tough fibrous tissue connecting bones together.

M

Magnesium Required by plants to make chlorophyll.

Malleable Can be hammered into shape or rolled into sheets.

Mineral Essential element required in small amounts for healthy animal or plant growth.

Mobile phase The liquid or gas used in chromatography which moves the solute.

Mole The number 6.02×10^{23}, which is the number of molecules in a the relative formula mass of a substance.

Molecule A particle made up of two or more atoms bonded together.

Moment The turning effect of a force about a pivot, calculated by multiplying the force by its perpendicular distance to the pivot. You measure a moment in units of Nm.

Monoculture The planting of the same crop year after year on the same piece of land.

Muscle Body tissue which contracts and relaxes.

N

Natural Originates from living material.

Neutralisation reaction The chemical reaction of an acid with a base in which they cancel each other out, forming a salt and water. If the base is a carbonate or hydrogen carbonate, carbon dioxide is also produced in the reaction.

Nitrate Soluble form of nitrogen, required by plants.

Nitrogen Required by plants to make DNA and amino acids.

Nucleus The very small and dense central part of an atom which contains protons and neutrons.

Nutrient Chemical needed by a living organism to maintain health or enable growth.

O

Organ A group of different tissues working together to carry out a particular function.

Organic A substance which contains (mainly) carbon in combination with other elements.

Organic farming Production of food and other materials without the use of chemicals. This method provides animals with an agreed standard of living.

Oxygen debt The extra oxygen that must be taken into the body after exercise has stopped to complete the aerobic respiration of lactic acid.

Oxygenated blood Blood that is rich in oxygen, on its way from the lungs.

P

Percentage yield The ratio of an actual yield, compared to the theoretical yield.

Pesticide A chemical used to kill pests.

PET Polyester.

Phosphate Soluble form of phosphorous, required by plants.

Phosphorus Required by plants for healthy roots.

Photosynthesis The process by which plants make food using carbon dioxide, water and light energy.

Physiotherapist Health professional who specialises in muscular and skeletal conditions.

Polarising microscope A microscope that uses polarised light to identify internal structure that cannot be recognised with ordinary light.

Polymer Organic compounds with long chain molecules made of many repeated bits linked together.

Potassium Required by plants for healthy leaves and flowers and a high fruit yield.

Precipitate An insoluble solid formed by a reaction taking place in solution.

Product A substance made as a result of a chemical reaction.

Protein A food group which provides the body with material for tissue repair, growth and some energy. A protein is a complex organic compound made from amino acids, and forms the basis of living tissue.

Q

Qualitative test Any way of detecting what substances are made of that does not involve calculations.

Quality control The monitoring of a manufactured product to ensure it meets a specification.

Quantitative test An analysis involving calculations, like a titration.

R

Radiate Energy being carried away by waves.

Reactant A substance we start with before a chemical reaction takes place.

Refractive index A measure of the light-bendability of a transparent substance, that depends on its density. Refractive index = $\sin i \div \sin r$.

Relative atomic mass The average mass of the atom of an element compared to carbon-12 (which has a mass of exactly 12).

Relative formula mass The total of the relative atomic masses, added up in the ratio shown in the chemical formula, of a substance.

Repeat A scientist can repeat a measurement if he obtains the same results from an investigation using same method and equipment.

Reproduce Results can be reproduced if the investigation is repeated by another person, or by using different equipment or techniques, and the same results are obtained.

Resolution The smallest size that can be measured.

Respiration The process by which food molecules are broken down to release energy for the cells.

Retention factor (R$_f$) The ratio of the distance travelled by a solute, to the distance travelled by the solvent.

Risk The likelihood that a hazard will actually cause harm and the seriousness of the consequence if it does.

Risk assessment An evaluation of the hazards in a standard procedure. We can reduce risk by identifying the hazard and doing something to protect against that hazard.

S

Saturated and unsaturated fats Saturated fats come from animal sources. Eating too much saturated fat can raise cholesterol levels leading to heart disease. Unsaturated fats come mainly from vegetables.

Scanning electron microscope Powerful electron-beam-based microscope that uses electrons to produce high-resolution images.

Scientific method The scientific method is a system of acquiring knowledge through observation and the experimental testing of hypotheses.

Selective breeding The process of selecting desired characteristics amongst organisms, through controlled breeding.

Serial dilution A microbiological technique, used to determine the number of bacteria present in a sample.

Solubility The mass of a substance (in grams) that will dissolve in 100 g of a solvent, such as water.

Solute A substance that is dissolved in a solvent to form a solution.

Solvent A liquid in which certain materials will dissolve.

Spring constant A measure of spring stiffness, calculated from force/extension.

Stain A dye or other coloring material, used in microscopy to make structures visible.

Standard procedure A clearly defined experimental method, used to ensure practical tasks are completed in a consistent manner.

Starch A source of carbohydrates obtained from cereals (like wheat and rice) and potatoes.

Stationary phase In chromatography, the medium through which the solute flows.

Stiff A stiff material is not flexible but stays rigid.

Streak plate An agar plate, on which a bacterial sample has been spread out (to identify individual colonies).

Stress When a force pulls on a material with a certain cross-sectional area, measured in units of force divided by area (N/mm^2). (Pressures push or squash, while stresses pull or strain.)

Strong A large force is needed to break a strong material.

Sugars A subgroup of carbohydrates. They dissolve in water and have a characteristic sweet taste.

Sweat gland A gland in the skin, which produces sweat to aid body cooling.

Synovial fluid A thick fluid released by the synovial membrane to lubricant joints.

Synthetic Man-made, rather than naturally occurring.

T

Tendon Inelastic tissue that connects muscle to bone.

Tensile strength The resistance of a material to pull it apart.

Tension A force that stretches something tight (rather than squashing or compressing it).

Theoretical yield The maximum amount of product a chemical reaction could produce.

Thermoplastic polymer A polymer that has few if any cross-links. Sometimes called a thermosoftening polymer as it is flexible and softens when heated, so it is easy to mould and shape.

Thermosetting polymer A polymer with strong cross-links, making it rigid once set. It cannot be remolded and decomposes, rather than melts, when heated.

Thorax The upper (chest) region of the body. In humans it includes the ribcage, heart and lungs.

Tidal volume (TV) The volume of air you breathe in and out normally.

Titration Technique in which the volume of liquid required for a chemical reaction is accurately measured.

Tough Absorbs energy when a forces applies stress. Tough materials are not brittle.

Trace evidence Evidence that exist in minute amounts, such as hairs, paint flecks, pollen, seeds, dust, soil, fibres and tiny flakes of glass.

V

Vein A blood vessel that carries blood towards the heart. (An artery takes blood away from the heart.)

Ventilation Breathing in (inhaling) and out (exhaling).

Ventricles The lower chambers of the heart. (See also Atria.)

Vital capacity (VC) The maximum volume of air you can force out of your lungs, after breathing in as hard as you can.

Vitamins and minerals Compounds needed in small amounts for normal growth of the body.

W

Water A liquid necessary for the life. Water molecules (H_2O) contain two hydrogen atoms and one oxygen atom.

Y

Yeast Single-celled fungi that can cause the fermentation of carbohydrates to produce carbon dioxide and ethanol (alcohol).

Yield An amount of chemical product.

Index

A

absorption 18
acid-base titration 116
acids 80
aqueous solutions 80
activation energy 89
actual yield 91
aerobic respiration 18
agar plates 72, 73
agricultural scientists 64, 65, 66, 67
agriculture 78–85, 94–5
alcohol 74, 75, 107, 110–11
algal blooms 84
alkalis 80
alloys 48–9
aluminium 45
alveoli 16
ammonia production 92–3
ammonium sulfate 81
amputee athletes 28–9
anaerobic respiration 19, 20, 74
analytical science 102–31
    modern instruments 122–3
    standard procedures 130–1
animals
    cells 124
    farming 82–3, 94
    selective breeding 94
antagonistic pairs 26
antibiotics 77, 82
antibodies 124
AQA marking criteria 137, 143
arteries 14
artificial joints 28–9
artificial skin 57
aseptic technique 72
Assignment 1
    marking grid 137
    maximising marks 138–9
    report-writing 138–9
    title 136
Assignment 2
    marking grid 143
    maximising marks 142–3
    practical activity evaluation 142
    workplace context 142
assignment report
    checklist 136, 141
    headings 141
    practical activity evaluation 141
    workplace context 141
athletes 28–9
atoms 48, 106, 114
atria 14
attraction, electrostatic 106

B

bacteria
    food poisoning 68, 70, 71
    food production 76–7
    growth 68–9, 71
    identification/measurement 72–3
balanced equations 114–15
base-line measurements 144
basic energy requirement (BER) 32
battery-farming 83
beer production 75
BER (basic energy requirement) 32
bicycle helmets 56
bicycles 56
biodiversity reduction 84–5
biological controls 79
biomaterials 57
biomechanics 26–7
bladder 23
blood 16, 124
    circulation 15
    group testing 124
    sugar levels 20–1, 36
BMI (body mass index) 33
Boardman, Chris 56
body mass index (BMI) 33
body movements 26–7
body temperature 22–3
bonding, chemicals 106–7
Borkenstein, Robert 110
bread making 74
breaking strength/stress 47, 58–9
breathalysers 110–11
breathing 16, 17
breathing rate 17
breeding, genetic 94–5
brewer's yeast 74
British Standards Institute (BSI) 44
brittle materials 47, 50, 51
BSI (British Standards Institute) 44
'bubble raft' model 49
bullets, forensic evidence 120

C

calcium oxide 90–1
calculations 115
Campylobacter bacteria 68
capillaries 22
carbohydrates 18, 35
carbon-ceramic composites 51
carbon dioxide extinguishers 7
carbon-fibre composites 50
cartilage 25
catalysts 89

[continued]

cattle breeding 94
CE marking 44
cells 124
cellulose 50
ceramics 50–1
cheese production 76–7
chemical analysis 106–7
    see also analytical science
chemical equations 114–15
chemical reactions
    endothermic 92
    exothermic 92
    rates 88–9
    reversible reactions 92–3
    yield 90–1, 93
chemicals
    farming use 78, 80–1
    production 81
chest 14
chicken farming 83
chromatography 118–19
chromium 49
cigarettes, forensic evidence 104
circulation of blood 15
clothing 52–3
cold climates, competing in 23
collision theory 88–9
comparison microscopes 120
composites 50, 51
compounds 106–7, 108
compression 46, 59
compressive breaking strength 59
concentration, chemical reactions 88
conduction 22, 45, 53
contrast, microscopy 120
control measures 5
controlled environments 79, 82
copper 45
core body temperature 22
corrosion resistance 43
covalent bonding 106, 107
crime scene evidence see forensic
    science
cross-links 54, 55

D

data collection/processing 141, 142
Defra (Department for Environment,
    Farming and Rural Affairs) 66, 104
density 44–5, 58
deoxygenated blood 16
deoxyribonucleic acid see DNA
Department for Environment, Farming
    and Rural Affairs (Defra) 66, 104
diabetes 20, 21, 36

diaphragm 16
diet 12, 30–1, 34–5, 64, 67
dieticians 64, 67
dislocations (metals) 49
DNA 124, 125
    database 125
    profiling 104, 126–7
    swabs 126
double pump (heart) 14
drink-driving 111
drinks, sports 34
drugs analysis 122–3
dry powder extinguishers 6, 7
dust analysis 121
dynamic equilibrium 92

E

E. coli bacteria 68
'ecological footprint' 85
egg production 83
elastic limit 46
electrical conductivity 45
electrical fires 7
electrolytes 34
electrons 48, 106
electrophoresis 126–7
electrostatic attraction 106
end point 116
endothermic reactions 92
energy
    activation energy 89
    body requirements 32–3
environment
    intensive farming impact 84–5
    pollution 84
    protection 104
    sampling 72
enzymes 74
EPS (expanded polystyrene) 56
Equations, balanced 114–15
erosion of soil 84
ethanol 74, 75, 107, 110–11
ethics 29
evaporation 22, 145
evidence
    collection/processing 141, 142
    scientists' use of 140–1
exercise
    changes during 18–19
    recovery after 20–1
exothermic reactions 92
expanded polystyrene (EPS) 56

F

farming 78–85, 94–5
fats 30
fermentation 74, 75, 77
fertilisers 79, 80, 81, 88
fibre (diet) 30

fibres, forensic analysis 121
fire doors 6
fire extinguishers 6–7
fire safety 6–7
fire triangle 6
fires, fatalities/injuries 6
fitness and health 12–37
flame tests 112–13
flammable materials 145
Flex-Foot 29
flexibility/stiffness test 59
fluid levels, body controls 22–3
foam extinguishers 7
food
    distribution cost 85
    energy levels 31
    hygiene 70–1
    labelling 31
'food miles' 85
food poisoning 68–9
food production 64–97
food science 64, 66–7
    standard procedures 96–7
Food Standards Agency (FSA) 66
foot and mouth disease 104
forces, on materials 46–7
foreign genes 94
forensic science 104
    blood group testing 124
    DNA evidence 104, 125, 126–7
    DNA profiling 104, 126–7
    drugs analysis 122–3
    electrophoresis 126–7
    flame tests 112–13
    glass analysis 128–9
    ion testing 108–9
    microscopic evidence 120–1
Forensic Science Service 122
formulae 107, 128
free-range egg production 83
frost-resistant tomatoes 95
fruit salad test 130
FSA (Food Standards Agency) 66
fuel cell breathalysers 111
fungicides 79

G

gas-liquid chromatography 122
gene pool 145
genes 94, 125
genetic breeding 94–5
genetic engineering 145
genetically modified crops 95
glass 56, 128–9
glossary 144–7
glucagon 20, 21
glucose 18, 30, 35, 36
glucose-testing strips 36
glycogen 20, 30, 35

H

Haber process 93
haemoglobin 124
hand-washing 105
'handgrip strength test' 36
hardness test 58
hazard symbols 2–3
hazards 2–3, 4
head, moments acting on 27
health and fitness 12–37
health and safety 5
Health and Safety Executive (HSE) 2
healthcare 13, 105
heart 14–15
heart rate (pulse) 15, 19
helmets 56
hens 83
herbicides 79
high-protein diets 35
hip replacements 28, 42, 48
Hooke's law 46–7
hot/humid climates, competing in 23
HSE (Health and Safety Executive) 2
hydration 34
hydroponics 86
hygiene 70–1, 105
hypertonic drinks 34, 35
hypothermia 23
hypotonic drinks 34

I

IAAF (International Association of
    Athletics Federations) 29
impact resistance 11
impact tests 51
information gathering, student reports
    138–9
infrared spectroscopy 111, 123
injuries 2, 6, 24, 28
insulation, thermal 53
insulin 20, 65
intensive farming 78, 79
    animal rearing 82–3
    chemicals use 80–1
    environmental impact 84–5
intercostal muscles 16
International Association of Athletics
    Federations (IAAF) 29
internet research 138–9
investigation report 136, 139
ionic bonding 106–7
ionic compounds 106–7, 108
ions 48, 106, 108–9
iron 43
isotonic drinks 34
Izod impact test 51

J

joints 25

K

Kevlar (material) 52–3
kidneys 23
Kitemarks 44
kiwi fruit 125

L

labelling of food 31
laboratories
    hazards 4–5
    safe working 3, 10
lactic acid 19, 20, 76, 77
laminated glass 56
level of water 22, 23
levers 26–7
ligaments 25
light, plant growth 87
lignin 50
liquid-gas chromatography 118
liver 20
Locard, Edmond 128
LotusSport Pursuit bicycle 56
lungs 16–17
Lycra 52

M

magnesium 80, 87
malleable materials 145
marking criteria 137, 143
mass, atomic 114
mass spectrometry 123
materials
    brittleness 47, 50, 51
    compressive strength 59
    density 44–5, 58
    electrical conductivity 45
    flexibility/stiffness test 59
    forces on 46–7
    hardness test 58
    measurement 44–5
    medical uses 57
    plastics 10
    properties 44–5
    shape 53
    sports uses 50, 56
    stiffness test 59
    stress 47
    tensile strength 58
    testing 10, 58–9
    toughness 47, 50, 51
    types 48
materials science 42–59
medical materials 57
medicines 105
metal ions 106, 109, 112

metals 48–9
    bonding 48
    dislocations 49
    models 48
    properties 48
microbiological techniques 72–3
microbiologists 64, 65, 67
microorganisms
    food poisoning 68, 70, 71
    food production 76–7
    growth 68–9, 71
    identification/measurement 72–3
microscale titration 117
microscopes 120–1
milk 71, 77
minerals 30
mobile solvents 118
mole (unit of substance) 114
molecules 107
moment (turning effect) 26–7
monocultures 85
moss, forensic analysis 120
muscle strength 36
muscles 16, 26, 27, 36

N

nanometres 120
National DNA Database 125
natural fibres 52
neutralisation reactions 80–1, 116
nitrates 80, 81, 87
nitrogen 145
non-metal ions 106, 109
NPK fertiliser 80
nutrients
    plant growth 87
    soil 79
nutrition 12, 30–1, 34–5, 67
nutritionists 67

O

oil immersion method, refraction 129
organic compounds 107
organic farming 78, 79, 82
organs 145
osteoarthritis 28
over-grazing 84
oxygen debt 19, 20
oxygenated blood 16

P

paint, forensic analysis 120, 121
pancreas 20
paper chromatography 118
Paralympics 28
pasteurisation 77
paternity tests 127
percentage yield 91

periodic table 106
personal hygiene 70, 71, 105
Perspex 128
pest control 79
pesticides 79, 84
PET (polyester) 56
pH scale 80
pharmaceuticals 105
phase change materials 53
phosphates 80, 81, 87
phosphorus 146
photosynthesis 78
physiology 12, 13
physiotherapy 13, 24–5
pig farming 82
Pistorius, Oscar 28, 29
planning, student assignments 141, 142
plant growth 78–9, 86–7
plasma 124
plastics
    polymers 28
    testing 10
    uses 54
platelets 124
poisoning
    food 68–9
    wildlife 84–5
polarising microscopes 120
pollen grains 121
pollution 84
polyester (PET) 56
polymers 28, 52, 54–5
positive ions 48
potassium 80, 81, 87
practical activity evaluation 141, 142
practical investigation 136, 140–3
precipitates 146
precipitation reactions 108–9
products 88
profiling, DNA 104, 126–7
prosthetics 28–9
proteins 30, 35
public health 105
pulse 15

Q

qualitative tests 96
quantitative tests 96

R

radiation 22, 23
reactants 88
reactions see chemical reactions
recipes 8
red blood cells 124
Redgrave, Steve 21
reference literature 104
refractive index 128–9
rehabilitation 24

relative atomic mass 114
relative formula mass 114, 115
rennet 76
report-writing 139
reproducible measurements 146
research methods, student reports 138–9
research report 136, 139
resolution, microscopy 120
respiration 18–19
resting heart rate 15, 19
retention factor equation 119
reversible reactions 92–3
risk assessment, assignments 141, 142
risks, safety 2–5
river pollution 84

S

safe working 2–11
safety signs 2–3
Salmonella bacteria 68
saturated fat 30
scanning electron microscopes (SEMs) 120
science
    practical techniques 136–43
    problem-solving 140–1
    use of evidence 136–7, 140–1
scientific method 8–9
selective breeding 94
SEMs (scanning electron microscopes) 120
serial dilution technique 73
shape, material properties 53, 57
shape memory materials 57
shoes, sport 45
signs, safety 2–3
silk 52
skeleton 12, 24
skin 22, 57
smoke alarms 6
sodium hydroxide test 109
soil
    erosion 84
    nutrients in 79
solubility 130
solutes 118
solutions, aqueous 80
solvents 107, 118
spaghetti hypothesis 55
spectrometry 123
spectroscopy 111
sports diets 35
sports drinks 34
sports injuries 24
sports materials 50, 56
sports scientists 13
sports shoes 45
springs 46–7
sprint shoes 45
stains, microscopy 120

stamina test 18
standard procedures
    analytical science 130–1
    carrying out 10–11
    following 8–9
    food science 96–7
    health and fitness 36–7
    materials testing 58–9
    microbiological techniques 72–3
    titration 116–17
standard solutions 116
standards 44
starch 18
stationary substances 118
steel 45, 47, 49
stents 57
sterilisation 72
stiffness test 59
streak plate technique 72–3
strength tests 36
stress 47
student investigation 136, 140–3
substance identification 103
substance unit (mole) 114
sugar
    blood levels 20–1, 36
    solubility test 130
sugars 18
surface area, chemical reactions 89
swabs, DNA analysis 126
sweating 22
symbol equations 114
symbols, hazard 2–3
synovial fluid 25
synovial joint 25
synthetic polymers 52, 54

T

tea 96
temperature
    body controls 22–3
    chemical reactions 88–9
tendons 25
tennis rackets 50
tensile strength 47, 58
tension 46, 47, 58
testing, materials 58–9
theoretical yield 90–1
thermal conductivity/insulation 53
thermoplastic polymers 54
thermosetting polymers 55
thin-layer chromatography (TLC) 118
thorax 14, 16
tidal volume (TV) 17
titanium 50
titrations 116–17
TLC (thin-layer chromatography) 118
tomatoes 95
tough materials 47, 50, 51
trace evidence 121

transport 56
tungsten 57
turning effects (moment) 26–7
TV (tidal volume) 17

U

UHT milk 71
ultra-heat treatment 71
unsaturated fat 30
urine 36

V

VC (vital capacity) 17
veins 14
ventilation (breathing) 16, 17
ventricles 14
vital capacity (VC) 17
vitamins 30

W

water
    blood levels 23
    body levels 22, 23
    in diet 30
    fire extinguishers 6, 7
    plant growth 87
websites 138
white blood cells 124
wildlife poisoning 84–5
windscreens 56
wine production 75
wood 50
word equations 114
working safely 2–11
workplace context, student assignments 141, 142

Y

yeast 74, 75
yield 90–1, 93
yoghurt production 65, 77

Z

zirconium 28

Photo acknowledgements

1.1.1 David Parker/Science Photo Library; 1.3.2 Comstock/Getty; 1.3.4 Nelson Thornes; 1.4.1 AJ Photo/Science Photo Library; 1.5.2 Jonathan A. Meyers/Science Photo Library.

2.0.1 Gustoimages/Science Photo Library; 2.0.2 Philippe Psaila/Science Photo Library; 2.0.3 Samuel Ashfield/Science Photo Library; 2.0.4 Mark Sykes/Science Photo Library; 2.0.5 Maximilian Stock Ltd/Science Photo Library; 2.0.6 Tony McConnell/Science Photo Library; 2.0.7 Samuel Ashfield/Science Photo Library; 2.0.8 Doncaster and Bassetlaw Hospitals/Science Photo Library; 2.0.9 iStockphoto; 2.1.1 Hybrid Medical Animations; 2.2.5 Mauro Fermariello/Science Photo Library; 2.3.1 Samuel Ashfield/Science Photo Library; 2.4.2 Ian Boddy/Science Photo Library; 2.4.4 Rex Features; 2.5.2 Edward Kinsman/Science Photo Library; 2.6.1 Science Photo Library; 2.6.2 Roger Harris/Science Photo Library; 2.6.3 Medical Images, Universal Images Group/Science Photo Library; 2.7.1 iStockphoto; 2.8.1 Zephyr/Science Photo Library; 2.8.2 Chris Hyde/Getty Images; 2.8.3 Duif du Toit/Gallo Images/Getty Images; 2.9.1 Photolibrary.com; 2.9.2 Karen Struthers/Fotolia; 2.11.1 Caro/Alamy; 2.11.2 Hugo Philpott/Reuters; 2.11.3 Sports Illustrated/Getty Images; 2.12.1 Saturn Stills/Science Photo Library.

3.0.1 W.T. Sullivan III/Science Photo Library; 3.0.2 Massimo Brega/Eurelios/Science Photo Library; 3.0.3 Brian Bell/Science Photo Library; 3.0.4 Peggy Greb/US Department of Agriculture/Science Photo Library; 3.0.5 Philippe Psaila/Science Photo Library; 3.0.6 Eye of Science/Science Photo Library; 3.0.7 Science Photo Library; 3.2.1 iStockphoto; 3.4.1 Brett Weinstein; 3.5.1 Yuji Sakai/Getty Images; 3.5.2 Cordelia Molloy/Science Photo Library; 3.6.1 illustrart/Fotolia; 3.7.1 Science & Society Picture Library/Getty Images; 3.7.2 Patrick Landmann/Science Photo Library.

4.0.1 iStockphoto; 4.0.2 Alexander Raths/Fotolia; 4.0.3 BSIP, Mendel/Science Photo Library; 4.0.4 iStockphoto; 4.0.5 iStockphoto; 4.0.6 iStockphoto; 4.0.7 iStockphoto; 4.1.1 Brian Bell/Science Photo Library; 4.1.2 Mauro Fermariello/Science Photo Library; 4.1.3 Geoff Tompkinson//Science Photo Library; 4.2.1 Science Photo Library; 4.2.2 iStockphoto; 4.2.3 Dr Gary Gaugler/Science Photo Library; 4.3.1 Volker Steger/Science Photo Library; 4.4.1 Sotiris Zafeiris/Science Photo Library; 4.4.4 Tek Image/Science Photo Library; 4.5.1 David Scharf/Science Photo Library; 4.5.2 Rosenfeld Images Ltd/Science Photo Library; 4.5.3 Ringwood Brewery; 4.6.2 Rosenfeld Images Ltd/Science Photo Library; 4.6.3 Stockbyte (NT); 4.6.4 Scimat/Science Photo Library; 4.7.1 Corel 603 (NT); 4.7.2 Nigel Cattlin/Holt Studios International/Science Photo Library; 4.8.1 David Aubrey/Science Photo Library; 4.8.2 Martyn F. Chillmaid/Science Photo Library; 4.9.1 Peter Menzel/Science Photo Library; 4.9.2 iStockphoto; 4.9.3 George Lepp/AGstockUSA/Science Photo Library; 4.9.4; Bryan Peterson/AGstockUSA/Science Photo Library; 4.10.1 Robert Brook/Science Photo Library; 4.10.2 Kaj R. Svensson/Science Photo Library; 4.11.2 Maximilian Stock Ltd/Science Photo Library; 4.11.3 NHPA/Joe Blossom; 4.11.4 Nigel Cattlin/FLPA; 4.11.5 Nigel Cattlin/FLPA; 4.11.6 Nigel Cattlin/FLPA; 4.13.1 Peter Bowater/Science Photo Library; 4.14.1 iStockphoto; p98 Nigel Cattlin/FLPA; p99 Alexander Raths/Fotolia; p100 Visuals Unlimited, Inc. Nigel Cattlin/Getty Images.

5.0.1 Mauro Fermariello/Science Photo Library; 5.0.2 Massimo Brega/The Lighthouse/Science Photo Library; 5.0.3 CC Studio/Science Photo Library; 5.0.4 Volker Steger/Science Photo Library; 5.0.5 Mason Morfit, Peter Arnold inc./Science Photo Library; 5.0.6 Mehau Kulyk/Science Photo Library; 5.0.7 Paul Rapson/Science Photo Library; 5.0.9 Grant Harrington/US Department of Agriculture/Science Photo Library; 5.0.10 Andrew Brookes, National Physical Laboratory/Science Photo Library; 5.1.1 Tek Image/Science Photo Library; 5.1.2 David Hay Jones/Science Photo Library; 5.1.3 Kevin Beebe, Custom Medical Stock Photo/Science Photo Library; 5.1.4 Chagnon/Science Photo Library; 5.2.1 Laguna Design/Science Photo Library; 5.3.1 James Holmes/Thomson Laboratories/Science Photo Library; 5.3.2 Shout/Rex Features; 5.4.1 Andrew Lambert Photography/Science Photo Library; 5.4.3 Michael Donne/Science Photo Library; 5.5.1a, b, d Andrew Lambert Photography/Science Photo Library; 5.5.1c David Taylor/Science Photo Library; 5.8.1a, b Andrew Lambert Photography/Science Photo Library; 5.9.1 Pascal Goetgheluck/Science Photo Library; 5.9.2 Mauro Fermariello/Science Photo Library; 5.9.3a Mauro Fermariello/Science Photo Library; 5.9.3b Dr Jeremy Burgess/Science Photo Library; 5.9.3c Volker Steger, Peter Arnold inc./Science Photo Library; 5.9.4 i Dr Klaus Boller/Science Photo Library; 5.9.4 ii, iv Eye of Science/Science Photo Library; 5.9.4 iii Dr Jeremy Burgess/Science Photo Library; 5.9 SQ1 Mauro Fermariello/Science Photo Library; 5.10.1 Dr Jurgen Scriba/Science Photo Library; 5.10.5 Tek Image/Science Photo Library; 5.11.1a Dr Gopal Murti/Science Photo Library; 5.11.1b Gunilla Elam/Science Photo Library; 5.11.2 Mauro Fermariello/Science Photo Library; 5.11.3 Revy, ISM/Science Photo Library; 5.12.1 Tek Image/Science Photo Library; 5.12.3 Tek image/Science Photo Library; 5.12.4 Laguna Design/Science Photo Library; 5.13.1 Tek image/Science Photo Library; 5.13.3 Geoff Tompkinson/Science Photo Library.

6.1.1 David R. Frazier/Science Photo Library.

7.1.1 Andrew Lambert Photography/Science Photo Library; 7.2.1 Paul Burn/Getty Images.